Future-Oriented Technology Analysis

Cristiano Cagnin • Michael Keenan
Ron Johnston • Fabiana Scapolo • Rémi Barré
Editors

Future-Oriented Technology Analysis

Strategic Intelligence for an Innovative Economy

Springer

Cristiano Cagnin
European Commission Joint Research Centre
Institute for Prospective Technological Studies
c/Inca Garcilaso
s/n 41092 Seville
Spain
cristiano.cagnin@ec.europa.eu

Fabiana Scapolo
European Commission Joint Research Centre
SDME 10/78
1049 Brussels
Belgium
fabiana.scapolo@ec.europa.eu

Michael Keenan
The University of Manchester
Manchester Institute of Innovation
Research (PREST)
Manchester Business School
MBS Harold Hankins
Booth Street West
Manchester
M13 9PL
UK
michael.keenan@mbs.ac.uk

Rémi Barré
Ministère de l'Enseignement Supérieur
et de la Recherche
Direction Générale de la Recherche
et de L'Innovation
Départment des Etudes et de la Prospective
1 rue Descartes
75231 Paris cedex 05
France
remi.barre@cnam.fr

Ron Johnston
University of Sydney
Australian Centre for Innovation
Faculty of Engineering
J13 - Room 246, Link Building
Off Shepherd Street
Darlington, NSW 2006
Australia
rj@netserv.eng.usyd.edu.au

ISBN 978-3-540-68809-9 e-ISBN 978-3-540-68811-2
DOI: 10.1007/978-3-540-68811-2

Library of Congress Control Number: 2008929586

Cover Design: WMXDesign GmbH, Heidelberg, Germany

Printed on acid-free paper

9 8 7 6 5 4 3 2 1

springer.com

Contents

Contributors

Mario Albornoz
Argentinian National Research Council (CONICET),
Buenos Aires, Argentina

Rémi Barré
CNAM, Paris, France
Département des Etudes et de la Prospective, Ministère de l'Enseignement Supérieur et de la Recherche, Paris, France

Cristiano Cagnin
Joint Research Centre Institute for Prospective Technological Studies (JRC-IPTS)
Seville, Spain

Kerstin Cuhls
Fraunhofer Institute for Systems and Innovation Research (ISI)
Karlsruhe, Germany

Luke Georghiou
PREST, University of Manchester, England

Jennifer Cassingena Harper
Council for Science and Technology, Malta

Gaston Heimeriks
Dutch Advisory Council for Science and Technology Policy, Netherlands

Ron Johnston
Australian Centre for Innovation, University of Sydney, Australia

Michael Keenan
PREST, University of Manchester, England

Alan Porter
Technology Policy and Assessment Center, Georgia Tech, Atlanta, USA

Michael Rader
Institute for Technology Assessment and Systems Analysis (ITAS),
Karlsruhe, Germany

Fabiana Scapolo
Joint Research Centre (JRC), Brussels, Belgium

Peter De Smedt
DG Research, European Commission,Brussels, Belgium

Matt Staton
A Bigger Splash, Barcelona, Spain

Philine Warnke
Fraunhofer Institute for Systems and Innovation Research (ISI), Karlsruhe, Germany

Chapter 1
Positioning Future-Oriented Technology Analysis

C. Cagnin and M. Keenan

1.1 Background

Over the last half century or so, various tools and techniques have been developed that seek to better anticipate and shape future technological developments. Some of these approaches, particularly early on, tended to be techno-determinist in their outlook, but more recently, a greater acknowledgement of the co-evolution of technology and society has led to the adoption of necessarily more complex perspectives. Some approaches have been purely quantitative, others purely qualitative, whilst a mix is often preferred. Some have involved only 'experts', whilst others have sought to initiate a societal dialogue. And some have sought to explore possible futures through extrapolation, whilst others have adopted a more normative stance, identifying targets and setting out action plans for achieving more desirable futures.

A variety of epistemic communities have grown up around anticipating (and, in some cases, shaping) technological futures. Perhaps the best known are the technology forecasters, but there are also other long-established communities around technology assessment, not to mention the broader field of futures studies. More recently, a technology foresight community has developed, which has its roots in the innovation studies field (see Chap. 2 in this volume and Miles 2008). There are some differences between all of these communities – in terms of their roots, practices, and knowledge claims – but there are a far greater number of similarities (see Chap. 3). Indeed, differences within communities are often greater than the differences between them, whilst many individual practitioners clearly transcend different traditions. With this in mind, the European Commission's (EC) Joint Research Centre Institute for Prospective Technological Studies (JRC-IPTS) has sought to begin a dialogue between these overlapping communities, using the label *Future-oriented Technology Analysis* (FTA) as a common umbrella term for technology foresight, technology forecasting and technology assessment. Through a series of JRC-IPTS sponsored biennial seminars, these communities have come together to exchange experiences and knowledge. These seminars have culminated in the

C. Cagnin et al. (eds.) *Future-Oriented Technology Analysis*,
doi:10.1007/978-3-540-68811-2, © Springer-Verlag Berlin Heidelberg 2008

publication of several journal special editions[1] and the production of this book.

As a result of this activity, some progress has been made – through mutual learning and exchange – towards the development of a shared understanding, though some work remains to be done before a fully-fledged FTA community can be said to have emerged. This is most obviously reflected in the bias of many of the contributions in this volume towards technology foresight, with fewer contributions on forecasting and technology assessment. This may lead the reader to equate FTA with technology foresight, but this would be a mistake. If anything, the technology forecasting and technology assessment communities are more mature and developed than technology foresight and have much to offer to the development of a more broad-based FTA. At the same time, the ever-growing popularity of (technology) foresight offers new opportunities for experimentation and innovation and the chance to employ FTA in more varied settings. Increased knowledge-sharing and even collaboration between different communities could therefore benefit all concerned.

In the remainder of this introductory chapter, we first outline work done around tracing the evolution of FTA and its essential characteristics. We then discuss the relationship between FTA and decision-making processes, an area where there exists much misunderstanding and misplaced expectations. Thereafter, we explicate some of the main challenges facing contemporary FTA, particularly around impacts and their assessment, while in a final section, we provide a summary of the chapters that follow.

1.2 Generations and Principles

To understand present-day FTA it is important to recognise that many of the successful approaches and methods which are still widely used today were developed during the 1950s and 1960s, undoubtedly influenced by the context of the Cold War. During the 1970s there was an extension of frameworks used mainly to better understand and shape technology developments into better understanding social needs and what society might expect of science and technology (S&T) in fulfilling their expectations as well as how to develop in such a context. This has anchored FTA, up to the present, firmly in the relation between science and technology on the one hand, and social needs on the other. According to Loveridge (2001), it was also during the 1970s that the limitations of traditional planning within industry were recognised, particularly in the light of major unpredicted events, such as the

[1] For example, see special editions of *Technological Forecasting and Social Change* vol. 75(4) and *Technological Analysis and Strategic Management* vol. 20(3) as well as *Technological Forecasting and Social Change*, vol. 72(9).

1973 oil crisis. Since then, there has been a substantial shift away from the apparent certainties of the Cartesian era of modelling and management, towards more contingent approaches, in both business and the public sector.[2] Indeed, most FTA practitioners today acknowledge and take into account the co-evolution of S&T and society in their work.

These shifts in FTA approach have been characterised by a number of authors as moving through successive generations or phases (Johnston 2002,2007; Cuhls 2003; Georghiou 2001,2007). Perhaps the best known of these generation models is the one proposed by Georghiou for technology foresight, where he identifies five generations. In the first generation, the focus is on forecasting of technological developments or the internal dynamics of technology, with ownership in the hands of experts. In the second generation, the focus is on the interplay of technology and markets. Technological development is understood in relation to its contribution to and influence from markets, and participation happens across the academic-industrial nexus. In the third generation, the market perspective is enhanced by inclusion of a broader social dimension, involving the concerns and inputs of social actors, and with a user-oriented (i.e. customer) perspective. The methods used and the knowledge base drawn upon are expanded to deal with issues concerning social trends and alternative institutional arrangements. In the fourth generation, foresight exercises have a distributed role in the science and innovation system, and often multiple organisations carry on exercises that are specific to their own needs, but which are coordinated with other activities. Finally, in the fifth generation there is a mix of foresight exercises which are distributed across many sites, and the concern of these activities is either on structures or actors within the STI system, or on the scientific/technological dimensions of broader social and economic issues. It is important to highlight that these generations are ideal types and that in practice they are not mutually exclusive. Indeed, it is not uncommon for exercises to exhibit characteristics of more than one generation. Nevertheless, many practitioners have moved towards the more recent of these generations in their activities.

Perhaps in contrast to such generational models, other authors have attempted to distil the essence of FTA (e.g. Gavigan et al. 2001; Keenan and Popper 2007), which has seen the explication of principles, as shown in Box 1.1. Such principles can be used to distinguish FTA from other decision-support techniques and can provide novice practitioners with a checklist of essential characteristics that their FTA activities should aim to have. Prominently emphasised is the future-orientation of FTA, as are the principles of participation and action-orientation. The latter is a particularly important consideration, with much attention paid to how FTA should relate to decision-making processes.

[2] For more on historical reference see: Loveridge (2001), Cagnin and Scapolo (2007) and Georghiou (2007) as well as Chap. 2 of this book.

Box 1.1 FTA Principles (Keenan and Popper 2007)

Principle of future-orientation: FTA is a future-oriented activity, though not in a predictive sense. In fact, FTA assumes that the future is not pre-determined, but can evolve in different directions, depending upon the actions of various players and the decisions taken today. In other words, the future can be actively shaped, at least to some extent, and there is a certain degree of freedom to choose among alternative, plausible futures, and hence to increase the likelihood of arriving at a preferred (selected) future state.

Principle of participation: FTA values the multiplicity of perspectives, interests, and knowledge held across a dispersed landscape of actors, and seeks to bring these together in processes of deliberation, analysis, and synthesis. Thus, FTA is not the preserve of a small group of experts or academics but involves a wider number of different groups of actors concerned with the issues at stake. Moreover, the results of FTA often have implications for a wide variety of actors, so it is important to involve these as far as possible throughout the process.[3]

Principle of evidence: FTA relies upon informed opinion and interpretation, as well as creative approaches in formulating conjectures on the future. However, these are seldom sufficient on their own and are complemented with various sorts of data from trend analyses and forecasting, bibliometrics, and official statistics, among other sources. Clearly, the future cannot be known with certainty and it is impossible to test conjectures on the future in the same way as one might test scientific knowledge claims. However, the plausibility of conjectures – as well as the original insights that they bring – are essentially 'market tested' by the decision-makers who rely upon such information. If they are to be convinced of the worth of FTA, then results should be based upon a sound knowledge base.

Principle of multidisciplinarity: FTA recognises that many of the problems we face today cannot be understood from a single perspective nor the solutions foundwithin a single discipline. Accordingly, FTA intentionally seeks to transcend traditional epistemic boundaries, bringing together different

[3] Daheim and Uerz (2006) have coined the term "open foresight", which is strongly linked with the concept of open innovation (Georghiou 2007). This refers to the involvement of relevant stakeholders, both from inside and outside the target organisation, hence promoting networking (Martin and Johnston 1999) and acting as a means of disturbance for the organisation. Thus, according to Georghiou (2007), FTA approaches should also be used to bring together not only those responsible for the development of the technological or other knowledge needed for innovation, but also those who are likely to make use of the technology or to provide the regulatory environment in which it develops. Moreover, it is extremely important that senior management within firms or policy makers feel ownership of FTA results through direct engagement. Therefore, involvement and engagement of key personalities in positions of influence, both in firms and in government, is key to enable FTA approaches to attain the expected impacts and benefits in the policy and decision making system.

(continued)

Box 1.1 (continued)

disciplines in processes of deliberation that result in improved understanding and new working relationships.

Principle of coordination: FTA enrols multiple actors to participate in decision arenas where conjectures on the future are contested and debated. Supported by various data and opinion, the FTA process aligns participant actors around emergent agendas, resulting in a coordinated mobilisation of people and resources.

Principle of action orientation: FTA is not only about analysing or contemplating future developments but supporting actors to actively shape the future. Therefore, FTA activities should only be undertaken when it is possible to use act on the results.

1.3 FTA and Decision-Making Processes[4]

Policy and strategy development are increasingly being interpreted as a continuous reflexive learning process that underlines the need for 'systemic instruments' (Smits and Kuhlmann 2004) to complement traditional steering approaches. FTA has the potential to offer such a set of systemic instruments, although there is still much debate around its interface with the policy process. Focusing upon technology foresight, a recent debate at the JRC-IPTS on the functions and benefits that foresight might have in the policy making system has given rise to the idea that there may in fact be two modes of foresight,[5] and perhaps similar modes apply to other FTA approaches:

- In 'mode 1' foresight, the objective is to improve and optimise the existing system, even if the process somehow pushes at boundaries through gradual evolution and incremental changes. Accordingly, policy and decision makers can easily become partners of the process because they have much to win from a more efficient system. The foresight process itself can be adapted to suit particular policy conditions and requirements (Weber 2006; Eriksson and Weber 2006; Havas et al. 2007).
- In 'mode 2' foresight, the aim is to debate and promote fundamental changes of established paradigms. This applies when the current system is perceived to be

[4] This section builds upon work carried out as part of the FORERA (Foresight for the European Research Area) Action within the JRC-IPTS, specifically around mutual learning workshops organised as part of the ForLearn project. For further information, see Da Costa et al. (2007) or visit the website: http://forlearn.jrc.es/guide/0_home/index.htm

[5] This is the summary of part of the results of the debate which took place in the last of a series of four mutual learning workshops, or consolidation workshop, focusing on the impacts of foresight on the policy making system. For more information see http://forlearn.jrc.es/guide/0_home/index.htm

fundamentally unsustainable and thus it becomes necessary to transcend it and to build a new system based upon different conditions and assumptions. Thus, 'mode 2' foresight is about questioning the existing system, initiating disruption, undermining existing world views, and raising the spectre of the incredible. Within this mode, it may be more important to highlight discrepancies than to emphasise consensus. As far as policy and decision making are concerned, one of the most important characteristics of 'mode 2' foresight is that decision makers are unlikely to control the process. Indeed, they may have much (or perceive to have much) to lose within a redefined system and might, therefore, become fierce opponents of such foresight exercises.

In other work carried out by JRC-IPTS, four basic types of structured (stakeholder) dialogue that characterise an ideal FTA exercise have been identified, each of which can be considered a different stage that has to be shaped and tailored in such a way as to attain expected impacts in the relevant policy and decision making system. These stages are as follows: (1) understanding the current situation, (2) exploring what could happen, (3) debating what stakeholders or participants would like to happen, and (4) deciding what should be done.

In the *first stage* – understanding the current situation – FTA approaches produce a number of insights about the future, such as the dynamics of change, new perspectives on the future, an understanding of future risks and opportunities, the definition of possible strategic options, a comprehension of system capabilities, an appreciation of the views of different stakeholders, etc. The anticipatory intelligence that results can improve the knowledge base of decision-making conceptualisation and design.

The *second stage* – exploring what could happen – is one of projection and exploration, carried out in 'hybrid fora' of actors who may have few opportunities to exchange views and may even hold opposing interests. Through a collective dialogue around the future, different interest groups can develop a shared understanding of the current situation, of the issues at stake, and of future challenges. Furthermore, individuals participating in the process can develop more "future-oriented" attitudes, and therefore make better informed choices and be ready to better accept and encourage changes going in the direction of any emerging shared vision.

In the *third stage* – debating what stakeholders or participants would like to have happen – FTA can contribute to an improved mode of governance in multi-layered and multi-actor decision making arenas. This stage improves transparency and can provide a legitimacy and efficiency to the decision making process, thereby increasing the acceptance and credibility of decisions (Martin and Johnston 1999).

The *fourth stage* – deciding what should be done – is not always tackled within the FTA process, and some practitioners argue that it should not be part of an FTA exercise at all, but rather a distinct political stage. Nevertheless, anticipatory intelligence is not always easily translated into options for decision making, especially if it originates from a collective process. Moreover, in a participative process, decision makers might be reluctant to communicate their hidden agendas or their needs to participants. Therefore, the translation of anticipatory intelligence into options for decision making has to take place with those responsible for making such decisions or with the key personalities in positions of influence, both in firms and in government.

1.4 What We Would Like to Know About FTA but still Don't

Over the past decade, FTA activities have multiplied across a wide spectrum of settings and at different levels. Reflecting this rich diversity of contexts, FTA activities have assumed a range of labels (e.g. technology foresight, technology assessment, technological forecasting, horizon scanning, technology road-mapping, critical technologies), have multiple objectives and rationales, and have used different methodological designs. By extension, expectations of outcomes and impacts tend to be context-dependent, and vary from concerns with the take-up of FTA knowledge in policy and investment decision processes, through to organisational vision-building or the active inclusion of normally excluded groups in decision-making processes and fora. This variety might suggest that different objectives, methodologies, and expected impacts can somehow be related to different contexts and conditions.

However, at the current time, there is still little understanding of the relationships among these variables, leading to a situation where much reinvention occurs in many settings. It can be argued that there is a need for some stock-taking of FTA activities, with a view to identifying patterns of relations between these variables that could serve policy-making or decision-makers in their contexts. This would be extremely relevant in the context of industrialising countries as FTA approaches still hold the promise of bringing about many benefits in such a context.

The challenge therefore is to better elucidate the relations between FTA context, content and approach, with a view to exploring the possibility of designing activities that are fit for purpose. Whilst many FTA practitioners argue that 'recipe books' are not possible for FTA activities given the various contingencies at play in any particular context, it should be possible to demonstrate that, for example, some methodological approaches are better suited than others in certain situations. Alternatively, notions of 'systemic' and 'adaptive' FTA, where FTA activities are responsive to evolving environments, provide a different approach to the design challenge. The definition of general limits of FTA and how these might be relayed in the management of expectations surrounding such activities is also important in addressing such a challenge.

At the same time, pleas from sponsors of FTA activities for better accounts of demonstrable impacts are as old as FTA itself. Yet, little work has been done in this area, with most accounts of impacts confined to individual case study descriptions. Practitioners are inclined to contend that evaluating the impacts of FTA activities is difficult, on account of their 'behavioural additionality', their distribution across a system of actors, and delayed effects associated with the time horizons involved, to name but a few reasons. The evaluative demands of sponsors are also often dismissed as being ill-informed and therefore unreasonable, relying upon overly-narrow linear models of cause-effect that draw upon rational models of decision-making. On the other hand, from the sponsors' perspective, without better and fuller accounts of impacts, the future sponsorship of FTA activities (and certainly their wider diffusion and expansion) is rendered more difficult and places the whole activity under threat.

It is therefore reasonable to conclude that FTA practitioners need to pay greater attention to accounting for outcomes and impacts before their activities can ever be more mainstreamed. More and better accounts of impacts from case studies could help to increase our understanding of FTA and its effects, but will be insufficient on their own. There is now a need to submit FTA practices to interpretation of their significance by the relevant disciplines of the social sciences and humanities (SSH). In this regard, some work has already been done in closely related areas, such as programme evaluation, futures studies, planning, and the study of evidence-based policy and scientific advice regimes. Extension of concepts and theoretical insights from these areas might therefore prove fruitful. But it is likely that a wider examination of FTA is now required, drawing upon relevant SSH disciplines, such as epistemology, political science, sociology, economics, and management and organisation science. These will provide a variety of interpretative lenses that offer the possibility to expand our conceptualisation of FTA, which will in turn improve the prospects for evaluating processes and outcomes (for example, through the development of suitable indicators).

The implications of what we know and what we don't are played out in different policy settings and contexts around the world. But besides public policy processes, FTA is also widely applied in other areas of socio-economic life, such as business and higher education. Considering business first, there is evidence that an increasing number of firms, industrial associations and industry foundations are using FTA tools for a variety of reasons, including horizon scanning (e.g. of weak signals), strategy setting, development of corporate visions, portfolio analysis, and as an aid in the management of supply chains. The tools being used include technology road mapping, scenario planning, internal and external surveys, and visioning, among others. Whilst there are a few descriptions in the literature of this work, there has been little coverage of how FTA activities fit into the firm (i.e. their embeddedness), how they relate to (innovation) strategy, and the conditions for their impact (or otherwise). Furthermore, outside of the firm, many industrial associations have used FTA tools to provide future-oriented insights for their sectors and to build collaborative linkages among members. In some instances, the public sector (mostly national and regional governments) has sought to promote private sector use of FTA approaches, particularly in small and medium-sized enterprises and towards the further development of industrial clusters. Again, little of this activity has been reviewed and critically analysed. The challenge therefore is to examine the implementation and use of FTA approaches in (and for) the private sector emphasising the impacts of FTA activities, linking their analysis to actual practice, to theories of the firm, to the innovation strategy literature, and so on. The embedding of FTA tools and concepts in companies is, from the organisational point of view, as interesting for analysis as the implementation approaches (e.g. results) enacted by associations, foundations or the like.

Turning to another popular area for the application of FTA, i.e. higher education, universities are increasingly facing new challenges brought on by a number of major disruptive drivers. These include: globalisation and the accompanying mobility of students and scientists; the impacts of new technologies (e.g. the impacts of the internet on teaching); demographic change; increased competition and the need to do well in national and global rankings; the ongoing rapid expansion of the sector (more students and increasing demand for postgraduate education); demands for a greater emphasis upon problem-oriented interdisciplinary research; and a continuing reassessment of relationships with the private sector and the innovation-related knowledge economy agenda (e.g. through third stream activities), to name but a few. One response to the resulting rapid change and its associated uncertainty has been to use FTA tools at the institutional and sectoral levels. Typically, these have been used to extrapolate current trends and drivers into the future, to assess alternative futures, and to build visions in which strategy can be based. A lot of work already exists in this area, yet little is known of its impacts. Therefore, the challenge now is to focus upon the implementation and use of FTA tools and approaches in the Higher Education (HE) sector. For example, there is a need to better understand which FTA approaches are being implemented in institutional settings, and to examine how they have contributed to organisational prioritisation, strategy and vision-building. Alternatively, there is also the need to examine the many government and EC-funded FTA activities that have focused upon the future of the sector and assess their contribution to HE reform. It is also critical to consider the linkages between FTA for the sector, for example, such as that carried out by national governments and the European Commission, and those FTA activities carried out by HE institutions.

Given this overall state of affairs, the objectives of this book are twofold. Firstly, it sets out to provide a better understanding of the context of FTA, such as its historical development, the evolution of approaches and their benefits, possible typologies and rationales for FTA, and some of the current challenges being faced by FTA practitioners and users. Secondly, it aims to offer an overview of latest developments by outlining some important applications of FTA in the areas of technological development, policy-making, business strategy, and higher education, among others. In this regard, the book attempts to address some of the gaps in knowledge outlined above.

1.5 Summary of Chapters

The first part of the book is devoted to better framing and contextualising FTA. In a short opening contribution, Johnston (Chap. 2) offers a brief account of the origin of the FTA brand and its many overlapping fields of practice, with specific emphasis upon forecasting and foresight. Following this, Rader and Porter (Chap. 3) address some of the challenges in better elucidating the relations between FTA context, content and approaches in the perspective of achieving outcomes. They do this by offering and illustrating a framework to help practitioners and scholars consider the choice of methods for particular studies. Through the creation of quality criteria to assess FTA activities and linking these with different typologies of FTA and the range of methods used in such studies, they develop a “ten commandments” list of how to fit methods to FTA types in order to achieve expected outcomes, stressing all the time that FTA is not a singular activity with a “one size fits all” methodology. They also shed light on the differences and similarities among the many terms and approaches that constitute the umbrella term of FTA, such as technology forecasting, technology assessment, roadmapping, technology foresight, and foresight.

A lot of claims are made in the name of FTA, yet there is insufficient understanding of the mechanisms necessary for expected outcomes to be achieved. Focusing specifically on foresight, Barré and Keenan (Chap. 4) outline some of the common rationales and objectives offered for using FTA but suggest that expectance of their realisation is more a leap of faith than evidence-based. Whilst further empirical research would help practitioners and users to better understand the role and value-added of FTA, it will be insufficient on its own. Also required is a better conceptualisation of the ‘programme theories’ of FTA around which process and impact models can be built. This will necessitate a turn towards the social sciences and humanities (SSH), and the authors introduce concepts from four theoretical approaches that could inform FTA planning, the expectations around FTA impacts, and, ultimately, the evaluation of FTA.

In contrast to many of the contributions in this book, Staton (Chap. 5) challenges current trends in FTA that call for closer links to the present and a clearer identification of what difference such activities have made. He argues there is a place in the range of FTA activities for work that does not have to prove value for money in terms of present problems and that this would be to recapture the early spirit of foresight. This would also imply a change in policy expectations for FTA, away from mastering the future in the name of fairly narrow, often political and sectoral interests towards exploring a more ethical opening towards the future on behalf of all citizens. However, he remains sceptical as to whether the ‘dialogic’ process that marks much FTA, with its aim of creating a common basis for action to address future issues through dialogue and consensus, is able to deliver anything more than a ‘longish tomorrow’, understood through the lens of our current concerns and interests, our current language and policy practice. As an alternative, Staton argues that another version of the future is possible, coming out of a different, non-dialectical philosophical tradition. Drawing on the work of Derrida, he suggests the future is ‘monstrous’, unforeseeable and undecidable, that for which we are unprepared. A future that is not ‘monstrous’ has already become a predictable and programmable tomorrow and there is little need to use FTA to explore this. FTA’s unique value is to help us to welcome the ‘monstrous arrivant’, to accord hospitality to the absolutely foreign or strange, unacceptable or repulsive as most transforming ideas have been before domestication.

In the second part of the book, chapters are devoted to exploring FTA applications in various settings. They each describe a particular FTA facet and together offer an overall view of the latest developments in the field, linking these with the many challenges outlined above, and introducing in some cases future avenues for research and development. In the first contribution, Warnke and Heimeriks (Chap. 6) investigate the possibility of using FTA approaches as a systemic innovation policy instrument to support policy makers in influencing innovation trajectories according to social needs. This is done by exploring how foresight can exploit the results stemming from social science research on the social shaping of technology. To guide policy makers to better understand the implications of technological innovation on the wider socio-economic framework, they argue there is a need to clearly comprehend the interplay between technological and socio-economic change. To do so, they turn to science and technology studies to show the co-evolution process of technology and society, and propose ways for policy making to intervene in these developments, using FTA.

De Smedt (Chap. 7) explores some of the core problems around the effectiveness of FTA approaches (or, as he calls them, Strategic Intelligence tools) to support decision making and suggests ways to improve policy practice through the analysis of how such applications can effectively support decisions. The departure point is a thorough understanding of what contributes to decision failure: bounded rationality; neglecting internal change; stickiness and friction of information and knowledge; and mental models. This gives way to the statement that policies, rather than being constituted by order and rationality, are often characterised by constant paradoxes of uncertainty, interpretation, context meaning, power, volatility, compressed views of time and space, and partial information. Hence, the conclusion is that FTA applications need to be turned into the strategic behaviour and cycles of policy and social actors to become effective in supporting decision making. Three complementary perspectives on policy change are proposed to help with this: window of opportunity; clarity of purpose; and legitimacy of policy evidence.

The application of FTA to the world of business is taken up by Cuhls and Johnston (Chap. 8) who distinguish between FTA in business and FTA for business in terms of their objectives, the most commonly used methods and the results and impacts generated. They identify the main goals of FTA in business to be to position the firm as a responsible partner in society; to promote organisational change; and to foster innovation. FTA for business can include several different sorts of activities, for example, using results from national foresight activities for strategic planning purposes; industry associations either carrying out or facilitating the use of foresight results for its members; foundations providing information to society or to SMEs; and multi-client studies either financed by firms, the EC or ministries or associations to promote future developments. On a final note, the authors stress the need to improve corporate FTA methodological approaches by, for example, grounding applications in quantitative analysis and economical modelling.

Despite the many changes that universities have had to face over recent years, there is a continuing expectation that further changes are coming and hence an apparent need for FTA to help institutions and their stakeholders to move forward. Georghiou and Cassingena Harper (Chap. 9) survey some of this work, and consider FTA efforts in individual institutions, at the national level, and at the international level. This extends to the institutional and sectoral use of FTA in HE and the impacts generated on policy and decision-making. The chapter shows that methodological approaches tend to be rather simple and rely heavily on expert opinion. There are also commonalities in content of the FTA work that has been carried out in HE, with globalisation, competition and student consumerism, the rise of new agents and functions, demographic pressures, technical change, collaboration with industry, transdisciplinarity, and new funding patterns commonly appearing as common drivers of change. The chapter goes on to consider the impacts of FTA in the HE sector, highlighting the different sorts of possible effects, before finishing with discussion of some of the key challenges that lay ahead.

Albornoz (Chap. 10) takes up some of the arguments highlighted in earlier chapters and applies them to a Latin American context. He outlines the difficulties that Latin American countries have in using FTA approaches and the benefits that the application of FTA could bring about in less developed countries. To resolve this tension, he argues it will be necessary to develop more institutional and human technical capabilities, to build appropriate social structures, to network heterogeneous actors and adequate management resources, to enable better access to a wide range of databases, to foster technology innovation, as well as to have more governmental and business demand for this kind of study. Nevertheless, in spite of the many difficulties in applying FTA instruments in Latin America, there is a great deal of experience to be drawn upon, and the author describes some of this, going back to the 1970s. He shows that Latin America was one of the forerunner regions in applying FTA, but has since lost momentum and its position due to the isolated character of such applications and, most importantly, because many of these experiences have been discontinued or have had their recommendations ignored and, therefore, no real impact has been achieved. Hence, the promise of FTA as a powerful tool to be effectively exploited in identifying opportunities for the region and to function as an instrument to support planning and decision making for further development has yet to be fulfilled.

Porter and Scapolo (Chap. 11) provide an in depth analysis of new methodological developments with a view of offering a more comprehensive picture of how to best design FTA activities that are fit for purpose. Their main objective is therefore to scan the methods available from the range of domains constituting the FTA umbrella. Such a scan enables the design of a typology of 51 methods arranged in 9 families. Based on this they outline novel application of existing methods, the combination of techniques within a single FTA exercise, and new methodological developments. By doing so they reach the conclusion that three themes for FTA methods development deserve consideration: first, the use of advanced tools that help process, search, mine, organise, display and interpret electronic information. Second, the need for methods that deals with human judgments. And third, the exploitation of communication tools.

Finally, in a short coda, Keenan and Barré (Chap. 12) consider a number of practical steps that will be required for the FTA community to further develop and flourish. These are grouped under five headings, namely capacity building, increasing community linkages, raising awareness of FTA among potential users, preparing to address global problems, and evaluating and monitoring FTA.

References

Cagnin, C. and Scapolo, F. (eds) (2007) Technical Report on a Foresight Training Course, EUR 22737 EN

Cuhls, K. (2003) From Forecasting to Foresight Processes – New Participative Foresight Activities in Germany, Journal of Forecasting, Vol. 23, pp. 93–111

Da Costa, O.; Warnke, P.; Cagnin, C. and Scapolo, F. (2007) Five Functions of Foresight for Policy Making, Technology Analysis and Strategic Management (forthcoming)

Daheim, C. and Uerz, G. (2006) Corporate Foresight in Europe: From Trend-Based Logics to Open Foresight, Second International Seville Seminar on Future-Oriented Technology Analysis, Seville, 28–29 September

Eriksson, E. A. and Weber, M. (2006) Adaptive Foresight: Navigating the Complex Landscape of Policy Strategies, Second International Seville Seminar on Future-Oriented Technology Analysis, Seville, 28–29 September

Gavigan J, Scapolo F, Keenan M, Miles I, Farhi F, Lecoq D, Capriati M, and Di Bartolomeo, T (2001) A Practical Guide to Regional Foresight, Seville: European Commission Joint Research Centre, EUR20128

Georghiou, L (2001) Third Generation Foresight – Integrating the Socio-Economic Dimension, International Conference on Technology Foresight – The approach to and the potential for New Technology Foresight, Science and Technology Foresight Center, National Institute of Science and Technology Policy (NISTEP), Ministry of Education, Culture, Sports, Science and Technology, Japan

Georghiou, L (2007) Future of Foresighting for Economic Development, UNIDO Technology Foresight Summit 2007, Budapest, 27–29 September

Havas, A., Schartinger, D and Weber, K. M. (2007) Experiences and Practices of Technology Foresight in the European Region, UNIDO Technology Foresight Summit 2007, Budapest, 27–29 September

Johnston, R. (2002) The State and Contribution of International Foresight: New Challenges – The Role of Foresight in the Selection of Research Policy Priorities, JRC-IPTS Seminar, Seville, 13–14 May

Johnston, R. (2007) Future Critical and Key Industrial Technologies as Driving Forces for Economic Development and Competitiveness, UNIDO Technology Foresight Summit 2007, Budapest, 27–29 September

Keenan M. and Popper R. (eds.) (2007) Guide to Research Infrastructures Foresight, EC, Brussels

Loveridge, D. J. (2001) Foresight, Course Book for the PREST Foresight Course, University of Manchester, Manchester

Martin, B. and Johnston, R. (1999) Technology Foresight for Wiring Up the National Innovation System. Experiences in Britain, Australia, and New Zealand, Technological Forecasting and Social Change Vol. 60, pp. 37–54

Miles, I. (2008) From Futures to Foresight, in Georghiou L, Cassingena Harper J, Keenan M, Miles I and Popper R (eds.) The Handbook of Technology Foresight, Edward Elgar, Cheltenham

Smits, R. and Kuhlmann, S. (2004) The Rise of Systemic Instruments in Innovation Policy, Foresight and Innovation Policy, Vol. 1, Nos.1/2, pp. 4.32

Weber, M. (2006) Foresight and Adaptive Planning as Complementary Elements in Anticipatory Policy-Making: A Conceptual and Methodological Approach, in VoB J.-P., Bauknecht, D. and Kemp, R. (eds.) Reflexive Governance for Sustainable Development, Edward Elgar, Cheltenham, 189–221

Part I
FTA in Context

Chapter 2
Historical Review of the Development of Future-Oriented Technology Analysis

R. Johnston

2.1 Introduction

The origin of the term 'future-oriented technology analysis' can be traced to the planning for the IPTS Seminar 'New Horizons and Challenges for Future-oriented Technology Analysis: New Technology Foresight, Forecasting and Assessment Methods' held in Seville, Spain in May 2004 (IPTS 2004). In the run-up to this seminar the Planning Committee prepared a preliminary paper to stimulate the potential participants to select topics on which to present papers and posters. This paper was entitled 'Technology Futures Analysis: Toward Integration of the Field and New Methods' (Technology Futures Analysis Methods Working Group, 2004). The paper introduced what it called an umbrella concept to encompass the wide variety of technology-oriented forecasting methods and practices, namely, Technology Futures Analysis (TFA). It is interesting that between that point and the actual seminar, a subtle, but crucial change took place in that TFA became FTA. The essence of that change was that 'technology-oriented' gave way to 'futures-oriented'. This indicated that the focus of the seminar would be clearly on the future and ways to develop useful information for shaping the future. The preparatory paper itself fostered the change in focus as it developed a series of challenging questions about the field of analysis of possible and desirable futures" (Scapolo 2005).

The objective of the Seminar was to analyse possible overlapping fields of practice among technology foresight, forecasting, intelligence, roadmapping, and assessment. The diversity among these disciplines reflects the complexity of demands for FTA relating to differences in scope (geographic scale and time horizon), relationship to decision making, the extent of participation, the purpose of the analysis (awareness raising, envisioning, consensus building, corporate technology planning, etc), and the reliability of source information.

To understand the power of the umbrella term of FTA, it is necessary to examine the nature and historical development of each of the component concepts.

C. Cagnin et al. (eds.) *Future-Oriented Technology Analysis*,
doi:10.1007/978-3-540-68811-2, © Springer-Verlag Berlin Heidelberg 2008

2.2 Forecasting and Foresight

The drive to understand and reveal the future is almost as ancient as human history and human inquiry. All cultures and civilizations have produced their prophets, seers, oracles, shamans or 'witch doctors', seeking insights through stars, animal entrails, cloud patterns, seasonal variations or hallucinogenic experiences. Indeed, it can be postulated that notions of past and future are an integral aspect of 'homo sapiens' and our own remarkable evolution.

There have been a number of historical analyses of the emergence and evolution of foresight and forecasting. Cuhls (2003) has provided a detailed account of the achievements and failures of technology forecasting over four decades. Georghiou (2001) proposed that the evolution of foresight could be characterised in terms of three successive generations, which has subsequently been extended to five generations (Georghiou 2007). Johnston (2002) proposed five stages in the chronology of foresight, with technology forecasting and futurism leading to technology foresight, from which emerged foresight, with its wider understanding of the economic and social processes that shape technology.

Looking back in more detail, it can be seen that it was the scientific progressives who lead the way, despairing of a world dominated by the horror of the First World War followed by the Depression, and who called for a new world order which looked to science and technology as a primary means of redemption. Prominent among them was H.G Wells, whose first major publication serialised in a magazine was subtitled, "An Experiment in Prophecy", He anticipated what the world would be like in the year 2000, with accurate predictions of modern transport resulting in the dispersion of the population from cities to suburbs, moral restrictions declining as men and women sought greater sexual freedom; and the formation of a European Union. He also argued, in a BBC broadcast in 1933, of the need for professors of foresight "we are surrounded by numerous professors of the past, but not one of the future" (Miles and Keenan 2003).

In the 1950s the US Department of Defence, looking at the development of new weapons systems, faced two specific needs:

> 'the need for a methodology to capture the reliable consensus of opinion of a large and diverse group of experts and the need to develop simulation models of future environments which would permit various policy alternatives and their consequences to be investigated'. (Bradfield et al. 2005)

The first led to the Delphi technique, the latter to systems analysis and scenario planning, developed within the RAND Corporation. However, it was the particular conditions of the 1960s that gave birth to the contemporary form and practice of foresight. The demonstrated effectiveness of operations research, leading to the growing influence of systems theory and thinking, together with the strategic challenges of the Cold War, provided a climate in which organised thinking about the future flourished. In contrast to the nineteenth century theories of social change, which dealt with large impersonal processes of evolution, the new approach was based on the deliberate intervention to direct change for specified ends.

In France, the Futuribles project was launched. In the UK, a 'Committee on the Next Thirty Years' was established. In the US, Herman Kahn left the RAND Corporation to establish the Hudson Institute, where he initiated a series of major studies on the future addressing economic and social policy issues, as opposed to his previous military focus. Daniel Bell (Kahn and Wiener 1967, p.xxv) attributed this emergence to the effects of economic recovery and growth:

> It arises from the simple fact that every society today is consciously committed to economic growth, to raising the standard of living of its people, and therefore to the planning, direction and control of social change. What makes the present studies, therefore, so completely different from those of the past is that they are oriented to specific social-policy purposes: and along with this new dimension, they are fashioned, self-consciously, by a new methodology that gives the promise of providing a more reliable foundation for realistic alternatives and choices, if not for exact prediction.

And the new methodology?

> We have begun to assemble statistical time-series both to plot trend lines and to extrapolate likely developments. The existence of a trend is no necessary guarantee that it will continue; but knowledge of trends and curves gives us more knowledge of likely developments. Along with time-series, we have begun to construct models or likely combinations of trends and developments in order to uncover the connections and causal relations between variables. And finally, with such simple techniques as the Delphi method, we seek to impose some controls by checking the informed guesses of one set of observers with those of others. (Kahn and Wiener, pp.xxvii–xxviii)

Kahn was appropriately prosaic in addressing the question of why we should speculate far ahead. Not because we could predict the future, but because:

> Such studies, even if only partially successful, contribute to interesting lectures, provocative teaching and stimulating conversation, all of which can broaden horizons and increase creativity – by no means negligible benefits. More important, these studies can affect basic beliefs, assumptions and emphases. Probably most important, is that long-range studies provide a context in which to do 5- and 10-year studies that can and do influence policy choices. (Kahn and Wiener, p.1)

While the initial focus was on public policy, it soon attracted the interest of the business community. Royal Dutch Shell initiated a 'Year 2000' study in 1967, which identified that the historical trajectory of year on year expansion could not continue, and that the oil industry faced a discontinuity. Pierre Wack, a planner at Shell Francaise, who was familiar with Kahn's work, proposed:

> To experiment with scenario planning as a potentially better framework for thinking about the future rather than continuing to rely on conventional forecasts which were likely to be wrong in the face of a discontinuity. The initial scenarios developed in 1971 ... proved extraordinarily successful in that they correctly identified an impending scarcity of oil and an ensuing pointed increase in oil prices; shortly thereafter scenario planning was extended throughout the company (Bradfield et al. 2005, p.798).

Companies were also included in the national forecast (today we would say foresight) activities in the USA (Gordon and Helmer 1964) and in Japan (Kagaku Gijutsuchô Keikakukyoku 1971). This was also the time when the first scenario approaches were tested (Kahn and Wiener 1968). It is reported (Bradfield et al.

2005, p.798) that General Electric also produced four alternative scenarios in 1971 of the global and US economic and socio-political conditions in 1980.

The growth in the application of foresight in the corporate sector was apparently quite strong during the 1970s and early 1980s, and is well-documented. A survey of US companies in 1981 found limited use prior to the oil crisis of 1974, but a substantial surge after that date, such that by the early 1980s, almost half of the US Fortune 1,000 industrial companies were actively using foresight techniques in their planning processes. The companies using foresight were characterised by their large size, planning horizons of more than 10 years, and capital intensity, as in the aerospace, chemicals and petroleum industries (Linneman and Klein 1979, 1983).

The pattern of adoption of foresight in planning was largely similar in European companies. Malaska (1985) Malaska et al. (1984) reported a period of experimental adoption of foresight techniques after 1973 and strong growth between 1976 and 1978, mainly on the part of large companies in capital intensive industries with long planning horizons such as petroleum, motor vehicles, and power supply.

The UNIDO Technology Foresight Manual (2005) reports:

> In the last two decades several large enterprises in such diverse sectors as energy, automotive, telecommunications and information technology have established foresight groups and strategic planning processes, which analyse the long-term prospects of new technologies and their impact on markets and corporate strategies. DaimlerChrysler's Society and Technology Research Group (STRG) is one of the first future research groups to be established within a company. Since 1979 it has investigated, in close cooperation with its customers, the factors shaping tomorrow's markets, technologies and products.

The use of foresight blossomed from the mid 1990s. Most OECD member countries (i.e. the advanced industrial nations), almost all European countries, and many Asian and South American countries have conducted national foresight studies (Johnston 2002). Over this period, there has been an increasing shift away from methodology-driven foresight studies, towards recognition of the variety of tools available to conduct foresight studies, each suitable for different purposes and with different strengths and limitations. The majority of these studies have been conducted at the national level. This reflects a stage in the development of the application of foresight to priority-setting and policy objectives. In general, these studies have been formulated and carried out by organisations with a national responsibility with regard to science and technology matters, be they a government department, or a semi-independent advisory body.

A detailed categorisation of national foresight studies has been made against the objectives pursued, which were identified as national competitiveness, vision building, identification of key or emerging technologies, creation of networks, information dissemination and education, and development of a forward-looking culture. It was concluded that "in countries where successive projects have been carried out, one can observe how the evolution in methods employed aims to increase the impact and effectiveness of foresight" (Gavigan and Scapolo 1999).

A particular form of technology forecasting went under the label of 'critical technologies', and was based on the assumption that certain technologies were key to future economic performance, and could be identified. In the US the driver was largely the fear of economic decline because of a superiority elsewhere, notably Japan, in developing new technologies. These were largely expert-based, and were conducted through the 1990s (Wagner and Popper 2003). Critical technologies foresights are performed elsewhere at national governmental level. The much emulated Japanese Delphi studies are oriented towards identifying critical technologies, although this is now being augmented with societal aspects. Japan has recently completed its eighth exercise of this kind; Korea, its third, and China and India their first (Johnston 2006).

A major promoter of foresight, and more broadly future-oriented technology analysis, has been the Institute for Prospective Technology Studies (IPTS) of the European Commission's Joint Research Centre (JRC, now Directorate General JRC of the EU Commission) located in Seville (Spain). Staff at IPTS have conducted many studies of future technologies and their impacts, and contributed to the development of more effective networks and practice among foresight practitioners.

The European Foresight Monitoring Network, established as a "foresight knowledge sharing platform" under the fifth Framework Program, to monitor foresight activity, now holds details of around 1800 separate foresight initiatives. Undoubtedly, this is an incomplete collection.

2.3 Technology Assessment

Technology Assessment had its origins largely in the United States in the 1970s, but was rapidly imitated and developed elsewhere. Its major objective was to assess the potential and implications of emerging and future technologies. The lead organisation was the US Office of Technology Assessment (OTA) which conducted a wide range of comprehensive, future oriented technology assessment exercises over the period 1974–1995. The OTA studies primarily served to inform Congressional interests as they considered legislative policy options. OTA studies were public and they served to inform the community about emerging technologies with unbiased, carefully reviewed analyses.

In Europe, technology assessment activities continue to flourish in some areas (the German-speaking countries, for example) while the term is practically defunct elsewhere. Presently, the term "technology assessment" is most closely associated with parliamentary activities. Several national and regional parliaments in Europe have their own semi-permanent capacities for TA, some created after the demise of the US OTA, and most recently the European Parliament has signed a framework contract with a group of these parliamentary TA units to provide services to its own panel with responsibility for Scientific Technology Options Assessment (STOA). There is a large, but largely fragmented FTA community in Europe, serving a broad variety of clients at various levels of government, including municipalities, regions, national governments and the various European Institutions.

There have been several attempts to form professional associations in the domain covered by TA. In the 1970s the International Society for Technology

Assessment flourished briefly; the International Association for Impact Assessment followed, and continues to be very active, but lately less involved with technology assessment. The European Society for Technology Assessment (ESTA) was linked with the regular ECTA (European Conferences on Technology Assessment) meetings of the early 1990s. The International Association of Technology Assessment and Forecasting Institutions (IATAFI) faded out of sight in the new millennium.

Most recently the "Netzwerk TA" has been created for the German-speaking countries and has held two conferences. In Germany, Netzwerk TA was preceded by a database and related activities on behalf of the Federal Ministry of Education and Research by Karlsruhe Research Centre's Institute for Technology Assessment and Systems Analysis. These covered the whole of Europe. Most relevant in this context is a still existing newsletter cum scientific journal, currently named "Technikfolgen Abschätzung – Theorie und Praxis" (Technology Assessment – Theory and Practice), which has a distinct knowledge-sharing function.

While there were plans to create some kind of umbrella activity under the label of "European Technology Assessment Network" (ETAN) from the mid-1990s on, the network as originally planned was never realised and the label ETAN was used for relatively small-scale activities.

A recent promising approach is that of the development of an 'early warning' or 'over the horizon' scanning capability in a number of countries, including the UK and Finland, designed to develop awareness and understanding of forthcoming science and technology and their implications. The studies for the STOA panel of the European Parliament have served to pinpoint critical aspects of technologies and their application which might require the attention of legislators at some later point in time.

2.4 Conclusion

With the emergence of the global knowledge economy and the increasing significance of access to knowledge as the basis of economic competitiveness, the importance of being able to 'peer into the future' has become recognised in steadily growing circles. This sets the scene for the Seminar on which this book is based, and for the development, refinement and impact of future-oriented technology analysis detailed in the following chapters.

References

Bradfield R, Wright G, Burt G, Cairns G, van der Heiden K (2005) The Origins and Evolution of Scenario Techniques in Long Range Business Planning, Futures, 37(8), pp 795–812

Cuhls K (2003) From Forecasting to Foresight Processes – New Participative Foresight Activities in Germany, Journal of Forecasting, 23, pp 93–111

European Foresight Monitoring Network, available at http://www.efmn.info/

Gavigan J, Scapolo F (1999) Matching Methods to the Mission; a Comparison of National Foresight Exercises, Foresight, 1, pp 495–517

Georghiou L (2001) Third Generation Foresight: Integrating the Socio-Economic Dimension. in Proceedings of the Second International Conference on Foresight, NISTEP, Japan

Georghiou, L (2007) Future of Foresighting for Economic Development, UNIDO Technology Foresight Summit 2007, Budapest, 27–29 September

Gordon TJ, Helmer O (1964) Report on a Long Range Forecasting Study, Rand Corporation, Santa Monica

IPTS (2004) Proceedings of the EU–US Scientific Seminar: 'New Horizons and Challenges for Future-oriented Technology Analysis: New Technology Foresight, Forecasting and Assessment Methods', Seville, Spain, IPTS reports are available at http://www.jrc.es/

Johnston R (2002) The State and Contribution of Foresight: New Challenges. in Proceedings of the Workshop on the Role of Foresight in the Selection of Research Policy Priorities' IPTS, Seville

Johnston R (2006) Technology Planning in Major Asian Countries: An Analysis of Recent Foresight Reports from China and India & Comparison with Japan and Korea, Report to the Department of Industry, accessible at http://www.aciic.org.au/?page=23

Kagaku Gijutsuchô Keikakukyoku (1971) Gijutsu Yosoku Hôkokusho (Technology Foresight Report, 1st Delphi), Tokyo

Kahn and Wiener (1967) The year 2000: A framework for speculation on the next thirty-three years, New York: Macmillan

Kahn H, Wiener (1968) Toward the Year 2000: Work in Progress, Boston, Beacon Press

Linneman R, Klein H (1979) The Use of Multiple Scenarios by US Industrial Companies, Long Range Planning,12(1), 83–90, cited in Bradfield et al

Linneman R, Klein H (1983) The Use of Multiple Scenarios by US Industrial Companies: A Comparison Study 1977–1981, Long Range Planning, 16(6), 94–101, cited in Bradfield et al

Malaska P (1985) Multiple Scenario Approach and Strategic Behaviour in European Companies, Strategic Management Journal, 6, 339–366, cited in Bradfield et al

Malaska P, Malmivirta M, Meristo T, Hanson S (1984) Scenarios in Europe: Who Uses them and Why? Long Range Planning, 17(5), 45–49, cited in Bradfield et al

Miles I, and Keenan M (2003) Two and a Half Cycles of Foresight in the UK, Theorie und Praxis, 12, 4149

Office of Technology Assessment reports are held at www.wws.princeton.edu/~ota/

Scapolo F (2005) New horizons and challenges for future-oriented technology analysis-the 2004 EU–US seminar, Technological Forecasting & Social Change, vol 72, pp 1059–1063

Technology Futures Analysis Methods Working Group (2004), [Porter AL, Ashton WB, Clar G, Coates JF, Cuhls K, Cunningham SW, Ducatel K, van der Duin P, Georghiou L, Gordon T, Linstone, H, Marchau V, Massari G, Miles I, Mogee M, Salo A, Scapolo F, Smits R, Thissen W], 'Technology Futures Analysis: Toward Integration of the Field and New Methods', Technological Forecasting and Social Change, 71, 287–303

UNIDO (2005) Technology Foresight Manual, Vienna, available at http://www.unido.org/data/forms.cfm?formcode=TFMANUALREG

Wagner CS and Popper SW (2003) Identifying Critical Technologies in the United States: a Review of the Federal Effort, Journal of Forecasting, 22(2/3), 113–128

Chapter 3
Fitting Future-Oriented Technology Analysis Methods to Study Types

M. Rader and A.L. Porter

3.1 Introduction

In June 2007, the European Parliament's Scientific Options Assessment (STOA) Panel celebrated its 20th anniversary with an exhibition at the Parliament's premises in Strasbourg. The opening ceremony included speeches by the Parliament's president, Hans-Gert Pöttering, and the European Commissioner responsible for Science and Research, Janez Potočnik.[1] His predecessor, Philippe Busquin is now the chairman of the STOA panel which since his assumption of the chair has had a framework contract with a group of technology assessment institutions working for national or regional parliaments in Europe.[2] This arrangement is leading to the production of a series of reports, based mainly on a review of existing literature and the involvement of experts. In particular M. Busquin, as chairman of the panel, has noted several times that such reports would also be of interest to national parliaments, notably those without their own capacities for technology assessment. In order to facilitate the necessary exchange, contacts have been established with the Directorate General Research of the European Commission, the organisation Busquin supervised for 5 years. While there is (no longer) any unit with explicit responsibility for technology assessment in DG Research, there is a Directorate with responsibility for, among other things, foresight. A meeting to sound out the feasibility of a Commission-funded project to disseminate findings for the STOA Panel included a discussion on the distinction between technology assessment and foresight, which finished with the conclusion that they overlap, although the two should not be totally confounded.

---

[1] See http://www.europarl.europa.eu/stoa/events/workshop/2007_experience/default_en.htm, accessed on 17 July 2007.

[2] The group is known as The European Technology Assessment Group (ETAG) and consists of the Danish Board of Technology, the Flemish Institute for Science and Technology Assessment, the Parliamentary Office of Science and Technology (UK), The Rathenau Institute (the Netherlands), led and coordinated by the Institute for technology Assessment and Systems Analysis of Karlsruhe Research Centre. See: http://www.itas.fzk.de/eng/etag/etag.htm

C. Cagnin et al. (eds.) *Future-Oriented Technology Analysis*,
doi:10.1007/978-3-540-68811-2,

On the one hand, the story illustrates the surprising kind of impact future-oriented technology studies can have; on the other hand, it illustrates the prevailing volatility. While activities within the Fifth Framework Programme of the Commission were under the heading of "technology assessment," the Sixth Framework Programme labelled similar activities "foresight," including some still explicitly labelled as "technology assessment". It is perhaps symptomatic of the pace of change that there was now a need to explain technology assessment to a group of high-level national officials with responsibility for foresight.

As in the case of STOA, some of the institutions active in future-oriented technology analysis have been in existence for considerable time and a look at their activities reveals a strong sense of continuity. However, subtle changes in terminology or to the names of institutions[3] indicate changing focus and expectations on the part of the clients. An analysis of the reasons behind such changes could in itself provide insights into the impact of FTA studies – one need only think of the derogatory "technology arrestment" – but we are here concerned mainly with the relationship between choice of approaches and methods and the impacts achieved by studies designed in different ways.

3.2 Features of FTA Activities

This paper looks at FTA types in conjunction with methods. It asks the question: which methods best serve different FTA aims? This prompts us to consider alternative analytical forms of future-oriented technology analyses:

- Technology intelligence
- Technology foresight
- Technology forecasting
- Technology roadmapping
- Technology assessment (TA)

Even such a list is not without its problems, since, as we will see, activities of certain kinds, like technology forecasting or technology roadmapping, can form part of another kind of study, such as technology foresight or technology assessment. Despite the lack of distinct definitions and a degree of interchangeability of labels, there are certain essential FTA features that affect types and methods:

- Apart from some limited scope technology forecasting, FTA activities seldom aim to predict "the" future, instead exploring possible futures of varying degrees of likelihood. Predictions on the availability of technology (technology forecasting,) are quite frequently made and used in broader FTA activities, but it is rather

[3] Such as STOA itself, which used to be called Scientific and Technological Options Assessment (panel) – it is now supposed to be scientifically assessing technological options.

more difficult to predict applications and diffusion of technology, so these topics are themselves frequently the subject of more exploratory analyses (c.f., Medina Vasquez 2006).

- In some cases, a specific future is regarded as desirable and an aim is to identify measures that could lead from the present state of affairs to this desirable future ("backcasting"). "Visions" like those developed by the German "Futur" activity or the NBIC initiative of the National Science Foundation and the US Department of Commerce serve a similar purpose, usually in a narrower sphere. Roadmapping pursues similar aims by defining milestones in the development of a technology and its applications.
- Technology assessment originally emerged with the aim of contributing to the balance of power between the legislative and executive branches of government, but has increasingly moved towards providing knowledge suitable for actively shaping technology. This has led to the emergence of such concepts as participatory technology assessment, constructive technology assessment, discursive TA, consensus conferences, etc.
- Foresight usually covers a broad range of technologies, increasingly also including the societal context of technology applications. The best known early technology foresight studies were on so-called "critical" technologies regarded as key to future economic development. While there has been an aversion to anything suggesting centralised national S&T planning in the US and, therefore, no national foresight, most technology foresight in other countries has been commissioned by national governments. On the other hand, in countries with a tradition of centralised planning, there is a danger that foresight will be misunderstood as a new tool of central planning (c.f., Böhle and Rader 2003, p.7).
- An important factor driving foresight activities has been globalisation and the attendant shift in the role of nation states. The identification of promising areas of science, research and technology likely to add to the attractiveness of certain locations for job creation has led, on the one hand, to stressing the network building functions of foresight – creating dialogues among the various actors and stakeholders with an interest in technology. On the other hand, we see some shift from the level of the nation state to more local levels, where regions or cities compete as attractive locations for research and its economic spin-offs.
- In all, foresight has shifted its focus away from the state to a broader range of stakeholders, including industry, resulting, to some extent, in a "convergence" of US FTA emphases, such as competitive technical intelligence and roadmapping, with those of Europe.
- Contrary to some popular perceptions, such as the distinction between TA, foresight, and technology forecasting made by the ESTO network (cf. Rader 2001, p. 4, but revised in Tübke et al. 2001), technology assessment can be focused either on a specific technology or group of technologies (technology-driven TA), or on technology-related problems (problem-driven TA). Since TA is frequently dealing with complex technological innovation issues beyond the control of the state, the results of TA studies are increasingly addressed to coalitions between

the state and societal actors, including experts, political and industrial decision makers, and stakeholders of all kinds (c.f., Petermann and Coenen 1999).

- A major distinction between foresight and technology assessment used to be the range of technology covered by exercises – the best known technology foresights address a broad range of technologies while technology assessments are narrowly focused. However, more recently foresight in countries which have conducted broader exercises has tended to focus more on specific cases. The Futur process in Germany is organised starting very broadly and narrowing the field as a result of consultation until it produces a limited number of "guiding visions" (Leitvisionen), intended to provide the framework for state S&T endeavours. After two cycles of classical foresight, the United Kingdom foresight programme was reshaped to focus on such specific topics as coastal protection and cognitive systems. The activities under these headings are difficult to distinguish from activities labelled "technology assessment" elsewhere.
- A further distinguishing feature between foresight and technology assessment is the time horizon, which is typically 30 or more years for foresight and rather shorter for technology assessment. The first Swedish foresight project pointed out the "Zeitgeist Problem" related to this aspect – the tendency to be captive to "the spirit of the times" and to assume that tomorrow's problems and visions will be very much the same as today's. This implies that the persons involved in the foresight tend to examine rather shorter ranged futures than hoped. An additional problem in this respect is that progress in some areas of technology is much faster than in others, so that foresight here is more difficult than in areas of slow movement.

Some of the questions regarding the relationship of FTA goals, methods, and outcomes emerging from this brief perspective are thus:

- How broad is the range of possible futures to be explored in the FTA study? How complex is the interaction of variables identifiable in the area?
- Is the goal to select a desired/desirable future state or to explore the implications of several possible future outcomes in order to be prepared and to take any necessary counter-measures should undesirable events and effects become visible?
- To what extent are the results expected to contribute towards shaping technology, its applications, and the context of technological development?
- Who is the primary client of the FTA, and which are the secondary addressees? What use is made of findings by the primary client? Are there other uses with significant impact, not originally foreseen or intended? How are these perceived by the primary client?
- How strong is the participatory element? What are its theoretical underpinnings and functions? Which stakeholders are involved? What is the overall impact of participation on the societal discourse on the issues at stake?
- What is the time horizon and degree of uncertainty for the FTA activity? What methods are applied to help avoid falling prey to the "Zeitgeist" trap or to help participants "think outside the box"?

- How timely are the findings with respect to regulatory needs, exploring technological options while these are available, public debate, etc.? Is the use of certain methods too time-consuming to produce timely outcomes?

3.3 Matching Goals and Methods

The aims of this chapter are to assemble information to guide selection of approaches and methods for FTA activities. While much experience exists with various kinds of FTA activity and practitioners learn from examples perceived to be successful, there have been few attempts at systematic evaluation of studies. Even the findings of the EC-funded TAMI project – which sought to create a structured dialogue between TA experts and decision-makers in the S&T policy area on the issues of method and impact assessment – are based mainly on discussion and selected cases. What are needed are reviews to match success in one or more respects with the FTA role, approach, and methods employed.

The European Foresight Monitoring Network (http://efmn.info) is operating a knowledge-sharing platform for foresight practitioners. This has published a series of foresight briefs and a series of annual mapping reports (e.g. Keenan et al. 2006). The latter publication is based on assessment of some 800 of over 1,400 collected foresight exercises, to give an indication of the popularity of combinations of methods with practitioners. Presumably popularity does, to a degree, reflect on success, but that is not thoroughly assessed.

A first question is how to measure success? While it is not unusual for FTA exercises to undergo peer review, such reviews seldom cover any measure of success in achieving goals such as informing decision-making, reception in the media or by the general public, or in creating networks of actors. Additionally, many FTA studies involve more than one scientific discipline, striving for multidisciplinarity, interdisciplinarity or transdisciplinarity. This creates problems of finding a common ground to face the subject, shared terminology, and agreement on pertinent findings. One criterion of success in such studies is how well the members of the disciplines involved are able to communicate and to cooperate.

Von Schomberg et al. (2005) have proposed a set of criteria to assess the quality of knowledge produced in foresight activities. This is essential to assess the validity of such knowledge, since foresight increasingly seeks to incorporate knowledge provided by members of different scientific disciplines and by non-experts with stakes in the technology. FTA assessment confronts lack of knowledge (or non-knowledge), and also uncertain or contested knowledge. Since decision-making frequently has to take place despite such uncertainties, we need quality standards and assessment methods. Proposed criteria are highlighted in Table 3.1 (detailed in the annex to the von Schomberg et al. 2005 paper).

Perceived quality of an FTA study is a prerequisite for its success – if approach, methods or their implementation are perceived as unscientific or otherwise flawed, decision-makers are unlikely to base their decisions on the results of the study. It is

Table 3.1 Technology foresight quality criteria

Issues	Dimension	Considerations			
Information pedigree	*Fitness for purpose*	Correspondence of information & issues: – Adequacy – Relevance	Accuracy	Comprehensiveness	
	Applicability	Access & availability	Intelligibility		
	Reliability	Control – Sources – Where from? – Sources – Method of generation – Verification – Consensus	Confidence – Peer acceptance – Legitimacy		
Quality of tools	*Fitness for purpose*	Adequacy/applicability	Relevance	Adaptability/flexibility	Transformation/ encoding
	Transparency	Model documentation	Sources of information	Arbitrariness – scientific set-ups	
	Legitimacy	Collegial consensus			
Information communicated through networking	*Intelligibility*	Compliance with target audience	Accessibility	Transparency	
	Communication of uncertainties	Recognition	Statement		
	Pedigree statement				
Information communicated from networking into the foresight exercise	*Acknowledgement of Input*				

possible that some actors or stakeholders will acknowledge a study while others reject it almost entirely. In such cases, the subject will most likely involve uncertain or lacking knowledge.

Available data, or information, interacts strongly with method suitability. Certain types of FTA are more conducive to empirical data availability and participatory mechanisms. Applying a particular method is constrained by the availability of the necessary data. Information is also partly a result of the methods employed and their data requirements. We can also question whether the methods result in the analytical information outputs required for the goals of the project. Is this information of sufficient quality? How transparent was the process by which the information was produced? Is the process itself regarded as legitimate by those affected in some way by its outcome? An important factor in the criteria proposed here is "networking" or participation, which refers both to the participants receiving information as intended and to their input to the study.

3.4 FTA Types

We have offered some criteria in considering the quality of FTA studies. However, if we are to develop guidelines for the selection of methods, we need typologies of FTA and of the repertoire of methods used in such studies. Others have recognized the diversity of contexts and approaches, and their implications for suitable processes and methods. For instance, Martin (1995) pointed out process opportunities to fulfil a range of foresight goals. Barré (2002), noting the rich variety of foresight activities, contrasts extension (divergence and stretching to consider fresh possibilities) and concentration (convergence and synthesis) processes. He offers a dual-axis typology that distinguishes: scale (how extensive) and intensiveness of learning efforts. This leads to a graphical depiction of foresight types, showing different relative emphases on analytics and participatory process. Salmenkaita and Salo (2004) compare industrial sector and governmental foresight activities, distinguishing an interesting triad of explicit, emergent, and embedded foresight variations. The TAMI project identified a non-exhaustive list of 21 specific roles that technology assessment has played in individual projects. Correspondingly, the project developed a typology of impacts, related to three issue dimensions: technological/scientific, societal, and political/policy oriented; and three impact (or goal) dimensions: increasing knowledge, forming attitudes or opinions, and initialising actions. The typology is actually the result of decomposing TA projects into individual steps, each with a distinct role and target. We suggest that "(t)he introduction of the *concept of roles reveals that TA plays more roles and has more impact than usually appreciated*…"(Decker and Ladikas 2004, p. 91, italics in original). The selected methods result from the issues that are the subject of the studies and from the roles the project has to play.

We offer dimensions that differentiate foresight forms, with implications for their conduct. These build on a paper by one of the authors (Porter, 2005, & forthcoming), which identifies nine dimensions that help categorize a given FTA activity (Table 3.2), enhanced by consideration of levels of complexity (Medina Vasquez, 2006).

This typology works by picking one value for each dimension. It could certainly be expanded or modified to suit situations. It's not hard to come up with additional state values for these dimensions or to add dimensions. To illustrate its application, let's consider two actual FTA studies – one American and one European.

The US National Reconnaissance Office (NRO) initiated an intriguing project in 1998 called "Proteus." It strove to develop truly fresh perspectives on intelligence needs and technologies to fulfil them. It did so using the scenario planning approach of a commercial facilitator, Deloitte Consulting. Focusing out to the year 2020, the project generated nine insights – i.e., fresh lenses different from Cold War themes. These provided new ways to consider (and then plan to address) issues in a changing world. Three workshops involving a range of intel-

Table 3.2 FTA studies typology

Issues	Dimension	State values			
Content	*Motivation*	Extrapolative	Normative		
	Drivers	Science (research)	Technology (Development)	Innovation	Context[4]
	Scope	Single topic or technology	Multiple technologies	Wide-ranging planning	Combinations[5]
	Locus	Institution	Sector	Nation/Region	Global
	Degree of uncertainty	Prediction (quite certain)	Forecast (estimable risk)	Foresight (highly uncertain)	Indeterminant (unknown)
	Time horizon	Short (1–2 years)	Mid-range (3–10 years)	Long (15+ years)	Monstrous[6]
	Purpose	Informational	Action-oriented		
Process	*Target users*	Few; knowledgeable	Diverse		
	Participation	Narrow mix, closed process	Intermediate	Diverse mix, representative process	
	Study duration	Day(s)	Month(s)	Year(s)	

[4] Blind (2006) discusses "regulatory foresight" as one interestingly different focus with clear value.

[5] As nice illustrations, Loikkanen et al. (2006) offer a framework to combine impact assessment and foresight in innovation policy analyses, and De Smedt considers rich interaction possibilities between foresight and decision processes in Chap. 7 of this volume.

[6] Staton (2006) makes the case for fresh, long range thinking not anchored in present realities.

ligence professionals and outsiders helped compose five scenarios – characterizations of the world of 2020 to stimulate consideration of issues and solutions. For instance, one was named "Amazon plague," wherein mutating viruses wrack the world, shrinking trade and the world's economy, with governments turning authoritarian or chaotic, and reliance on the global information grid in lieu of reduced physical interchanges. Follow-on stages aimed to transfer Proteus thinking to other agencies, implement gaming environments, and assess the potential of emerging technologies to contribute to multiple future environments (Krause, 2002).

Table 3.3 offers an outsider's assessment of the NRO FTA effort. The motivation for Proteus is extrapolative – it seeks to anticipate potential changing world contexts, not to project desired states of affairs (a normative approach). The exercise is driven, not by S&T advances per se, but by the socio-economic-environmental context evolution (or revolution as the case may be). This is global, long range, informational foresight. The process is quite diverse, but not broadly representative. The premise is that this characterization provides vital clues on how to perform the foresight exercise in question and what methods are more suitable. The creative bent of Proteus suggests some will find these foresight activities and/or outputs more palatable than will others.

For the sake of contrast, let's also consider a TA study, INDICARE (INformed DIalogue about Consumer Acceptability of DRM Solutions in Europe), which was concerned with Digital Rights Management Systems. Table 3.4 casually characterizes it, this time by someone who was involved in parts of the project.

Table 3.3 Typology applied to the Proteus Study (State values in bold)

Issues	Dimension	State values			
Content	*Motivation*	Extrapolative	**Normative**		
	Drivers	Science (research)	Technology (development)	Innovation	**Context**
	Scope	Single topic or technology	Multiple technologies	**Wide-ranging planning**	
	Locus	Institution	Sector	Nation/Region	**Global**
	Degree of uncertainty	Prediction (quite certain)	Forecast (estimable risk)	**Foresight (highly uncertain)**	Indeterminant (unknown)
	Time horizon	Short (1–2 years)	Mid-range (3–10 years)	**Long (15 + years)**	
	Purpose	Informational	Action-oriented		
Process	*Target users*	Few; knowledgeable	**Diverse**		
	Participation	Narrow mix, closed process	**Intermediate**	Diverse mix, representative process	
	Study duration	Day(s)	**Month(s)**	Year(s)	

Table 3.4 Typology applied to INDICARE (State values in bold)

Issues	Dimension	State values			
Content	*Motivation*	Extrapolative	**Normative**		
	Drivers	Science (research)	Technology (development)	**Innovation**	**Context**
	Scope	**Single topic or technology**	Multiple technologies	Wide-ranging planning	
	Locus	Institution	Sector	**Nation/Region**	Global
	Degree of uncertainty	Prediction (quite certain)	**Forecast (estimable risk)**	Foresight (highly uncertain)	Indeterminant (unknown)
	Time horizon	**Short to mid-range**	Mid-range (3–10 year)	Long (15+ years)	
Process	*Target users*	Few; knowl-edgeable	**Diverse**		
	Participation	Narrow mix, closed process	Intermediate	**Diverse mix, representative process**	
	Study duration	Day(s)	Month(s)	**Year(s)**	

The overall goal of INDICARE was "to raise awareness, help to reconcile heterogeneous interests of multiple players, and to support the emergence of a common European position with regard to consumer and user issues of Digital Rights Management (DRM) solutions".[7] The method used in INDICARE was to set up a discourse among the various actors with interests in the area concerned, which is currently the subject of great controversy, in particular between copyright owners (publishers, music and film companies) and consumers. A possible impact of unrestrained development could be widespread criminalisation of citizens illegally using material covered by intellectual property rights, or restrictions in use compared to the "old" situation. A part of the study was devoted to finding out what is acceptable to as many parties as possible, i.e. to explore normative implications. Since the problems are already visible and certain technological solutions already available, the study was focused on the short to medium time range. This is a technology policy analysis, in contrast to Proteus, a creative exploration of alternative futures. The project worked by setting up fora of actors to exchange their viewpoints, mainly on the internet, but also at meetings with a smaller number of experts and stakeholders. An important instrument was an on-line journal that contained articles on controversial topics with the aim of stimulating a debate. Examples for such topics were new legislation in different countries or rulings by courts of law on contested issues, such as private copying. Due to the project character of

[7] See www.indicare.org (accessed on 21 July 2006).

INDICARE, which resulted in financing for a limited period of time, INDICARE was able to show the efficacy of its method for the problem at hand, but did not contribute to solving all problems existent in the field. Its most important result, apart from showing the way, was probably to bring together actors with different interests in the field and to demonstrate to some of these stakeholders the need to consider consumer interests. To this extent, INDICARE might have helped to shape technology in a limited field.

The next section of this chapter puts this FTA typing to use in helping guide the selection of appropriate methods.

3.5 Putting the Pieces Together: FTA Methods and Types

Different types of FTA demand different methods. As per Table 3.2 and Table 11.1 described in Chap. 11, the types and methods are too complex to make a simple prescription of what to do, when. The main message is to reflect on the FTA study at hand to consider alternative methods (tools, processes, etc.), then, weigh the pro's and con's of different approaches.

Boehle et al. (2001) followed up on the use made of six studies focused on electronic payment systems, and the FISTERA project drew a number of methodological conclusions from a comparative analysis of eight national foresight studies from European countries (Rader et al. 2003; Rader and Boehle, 2003). We list some of the findings pertinent to selection of methods:

- FTA can play an important role in organising a dialogue among actors who might otherwise not be communicating with each other (Bohle et al. 2001).
- If FTA is to directly support political decision-making, projects should be specific and provided with sufficient resources (Boehle et al, 2001).
- Electronic communication tools find use for speed, to enable the participation of dispersed communities, to enable structured discussion etc.
- There is need for a clear sense of the timing of technological developments, a distinction between short-term problems and long-term structural change of society. FTA scenario techniques are useful for long-term analysis (Boehle et al. 2001).
- The success of methods does not only depend on "fitness for purpose", but also on the provision of adequate resources. It has been pointed out, for example, that the Delphi study in the first UK Foresight cycle was not adequately prepared. It was replaced in the second cycle by a resource termed "the knowledge pool." That contained general programme information, access to scenarios and views about the future, and management information and working notes for foresight panels (Miles and Keenan 2003). Although useful for actors familiar with foresight, the resource proved daunting for newcomers and overall failed due to its sophistication (Miles and Keenan 2003, p.45).
- Another common shortcoming in foresight studies was failure to ensure mutual consideration of other participants' work in studies comprising several panels or

streams considering technologies with areas of overlap, like the information and communication technologies (cf. Rader and Boehle 2003, p. 76).

- The analysis of large-scale national foresight studies for FISTERA indicates that the discussion and feedback element was apparently the most successful component of foresight studies, a contribution to what Martin and Johnston (1999) term "the wiring up" of innovation systems.
- Sophisticated methods, such as full-blown Delphi studies, seem to be more successful if they form the centrepiece of a study rather than one of a broader range of methods.

We offer some further observations on the foresight types and suitable methods. Considering the Type dimensions (Table 3.2):

- Motivation:
  - Normative studies warrant more emphasis on the prescriptive methods (i.e., Valuing/Decision-aiding/Economic.
- Drivers:
  - Science-centred foresight requires substantial rethinking of tools devised to forecasting more incremental technological development processes. It is more subject to drastic change – i.e., breakthroughs. This suggests inclusion of Creativity Approaches, with heavy emphasis on Monitoring & Intelligence and also the use of such instruments as "wild cards" – unlikely but not impossible developments. Rapid foresight also becomes essential to respond quickly to discoveries.
  - Innovation-oriented studies also differ from traditional technology forecasting. They demand more attention to socio-economic contextual forces interacting with emerging technical capabilities to effect commercial products and services. Competitive technical intelligence approaches come to bear. Description, Modelling, and Logical/Causal analyses of competitive environments are vital.
  - Studies driven by contextual factors shift the focus to non-technical influences, requiring different sorts of expertise. Methods such as Scenarios come prominently into play.
- Scope
  - Tighter foci enable more data-based analyses.
- Locus
  - Institution-oriented studies enable tailoring of issues. For instance, exploration of emerging technology opportunities can be crossed against the institution's relative purposes and strengths, using Matrices.
  - Expanding Locus interacts with Process dimensions importantly – i.e., Participation considerations and suitable means to achieve these change drastically from institutional to national or trans-national (e.g., European Union) loci.

- Time horizon
  - Suitable methods shift as the time frame stretches. For instance, trend analyses long-term become very unreliable.
- Purpose
  - Action-oriented foresight leans toward assessment of policy options. Creativity approaches can aid in identifying a wider range of alternatives to consider. More prescriptive methods can help expose the advantages and disadvantages of these.
- Target users
  - As the intended users become more varied and numerous, increased attention to effective communication is critical. As a generalization, we invest way too high a portion of our resources in analyses, with too little in communication. This applies in particular to foresight studies targeted at broad communities of stakeholders, the results of which are frequently intended to stimulate a public debate. Roadmapping and Scenarios may be particularly beneficial.
- Participation
  - More inclusive foresight processes exert pressures on which methods are apt to work well. Participants like to understand; "black boxes" don't go over well, so transparency is important. For instance, highly elaborate modelling is probably unsuitable unless it can be simply explained. Suitable information visualization techniques may help convey information and analyses.[8]
- Study duration
  - While we are not emphasizing "needed in a day" foresight exercises, quick response to queries and challenges can be helpful. For instance, in dialogues among foresight process participants and/or with foresight users, interactivity helps. Try to enable "What if" analyses, done real-time, so that someone can ask about an alternative, and in a minute have a simulation run to reflect it.

3.6 Conclusions

We consider future-oriented technology analyses as multi-dimensional activities. Thus, the conduct of FTA needs to be tailored to the type. We have drawn up an extensive list of methods, with some suggestions on factors to consider in fitting these to the type of FTA being undertaken. FTA is not a singular activity with "one size fits all" methodology.

---

[8] Steyaert, Eggermont and Vandebosch (2006) illustrate the use of play acting to enable active participation of lay persons.

We need to recognize the inherent limitations of foresight and other types of FTA. In a review of US foresight activities (Porter and Ashton 2008), there was a reflection on the pros and cons of the U.S. "anti-foresight" stance. US institutions certainly conduct many future-oriented technology analyses, but the country has a distaste for centralized R&D priority setting or innovation planning (Wagner and Popper 2003). What's to be said for an anti-foresight approach? The more an innovation system is subject to unpredictable, rapid changes, the more advantage to the pluralistic approach. Good technological intelligence to pick up quickly on emergent opportunities may outweigh careful foresight. As our emerging technologies become more science-based (e.g., biotech, nano), we need to rely more heavily on Creativity approaches and Monitoring & Intelligence; less on Trend Analyses. A messy, pluralistic (i.e., not heavily planned) approach may especially do better at seizing sudden opportunities. Pursuing "Radical Innovation" calls for less forecasting, per se, than does pursuing incremental innovation (Dismukes et al. 2005).

Box 3.1 spotlights some things to note about these FTA methods and their application – taking the "commandment" notion lightly. To elaborate a bit:

- It can be helpful to consider the triad of Data, Theory, and Methods (Porter et al. 1991). To analyze a given issue, suitable methods must be selected on the basis of data availability. The complexity of many socio-technical developments exceeds our causal/predictive theory, constraining our attempts at causal reasoning or other intricate methodology.
- Given the Data/Theory/Methods concerns, it is advisable to use multiple methods that counterbalance each other's weaknesses.

Box 3.1 Fit methods to FTA types to achieve goals: "Ten Commandments"

1. Focus on the triad of Data, Theory & Methods to exploit the data, knowledgeably & appropriately
2. Time available for a study constrains what methods will work
3. Use multiple methods
4. Blend quantitative + qualitative methods (e.g., empirical + expert opinion)
5. Integrate distinct methodological contributions + multiple human judgments
6. Assess FTA quality
7. Devise an explicit path to impact decision-making
8. Formulate an explicit communication plan, taking advantage of electronic media as suitable
9. Be prepared; some methods require TRAINING
10. RESEARCH on FTA methods needs support

- Study resources and the time available also need to be factored into determination of what methods to use. FTA results must be available in a timely manner or they will fail to achieve their primary goals (recall the U.S. OTA situation).
- These techniques reflect both qualitative and quantitative approaches; integrating these is usually desirable.
- Quality assessment within studies would benefit from explicit success measures put forth at the initiation of the FTA effort.
- Not emphasized in this consideration of FTA content and process types, and methods, is the essential need to determine and act upon plans to identify the target users and how to actively engage them in the study (Porter et al., 2004). Effective communication of FTA activities and implications demands attention from the start.
- Some of the methods are intuitive; others benefit from experience and training, or even require this to be successful. The methods toolkit is a work in progress; research on new methods is badly needed.

This chapter offers frameworks of FTA quality considerations, study types, and methods. It gingerly offers pointers toward which methods may suit a 'particular studies' types and goals. Nevertheless, it is important to highlight that FTA types and methods can look toward a dynamic future in their own right!

References

Barré, R. (2002): Synthesis of technology foresight, in A. Tubke, K. Ducatel, J-P. Gavigan, and P. Moncada-Paterno-Castello (Eds.), Strategic Policy Intelligence: Current Trends, the State of Play and Perspectives. Institute for Prospective Technological Studies (IPTS), Technical Report EUR-20137-EN, Seville

Blind, K. (2006): Regulatory Foresight: Methodologies and Selected Applications, FTA 2006 Seminar (Future-oriented Technology Analysis), Institute for Prospective Technological Studies, Seville, 28–29 September

Boehle, K., Rader, M. (2003). Foresight Somewhere in the Expanse Between Futures and Technology Assessment, Technikfolgenabschätzung – Theorie und Praxis, Vol. 12, No. 2, (June), pp. 5–10

Boehle, K., Rader, M., Riehm, U., Weber, A. (2001): Technology Assessment and Electronic Money – Between Consultancy and Oversight. Proceedings of Conference: Innovations for an e-Society – Challenges for Technology Assessment. 17–19 October 2001. Teltow: VDI-VDE, (CD-ROM)

Decker, M., Ladikas, M. (eds) (2004): Bridges Between Science, Society and Policy. Technology Assessment – Methods and Impacts, Springer, Berlin Heidelberg New York

De Smedt, P. (2006): Interactions Between Foresight and Decision-making, FTA 2006 Seminar (Future-oriented Technology Analysis), Institute for Prospective Technological Studies, Seville, 28–29 September

Dismukes, J.P., Miller, L.K., Bers, J.A., and McCreary, W.N. (2005), Technologies of Thinking Seen Key to Accelerated Radical Innovation, Research Technology Management, Vol. 48, pp. 2–4

Krause, P.H. (2002) The PROTEUS Project Scenario-Based Planning in a Unique Organization, Technological Forecasting and Social Change, Vol. 69, pp. 479–484

Keenan, M., Butter, M., Sainz de la Fuenta, G., and Popper, R. (2006): Mapping Foresight in Europe and Other Regions of the World: The 2006 Annual Mapping Report of the EFMN, European Foresight Monitoring Network (www.efmn.eu)

Loikkanen, T., Kutiniahti, P., and Eerola, A. (2006): Towards an Integrated Framework of Impact Assessment and Foresight Studies in Innovation Policy Analysis, FTA 2006 Seminar (Future-Oriented Technology Analysis), Institute for Prospective Technological Studies, Seville, 28–29 September

Martin, B. (1995): Foresight in Science and Technology, Technology Analysis & Strategic Management, Vol. 7, No. 2, pp. 139–168

Martin, B.R., Johnston, R. (1999): Technology Foresight for Wiring Up the National Innovation System, Technological Forecasting and Social Change, Vol. 60, pp. 37–54

Medina Vasquez, J. (2006): Map of Levels of Complexity and Indetermination for Foresight Studies, FTA 2006 Seminar (Future-Oriented Technology Analysis), Institute for Prospective Technological Studies, Seville, 28–29 September

Miles, I., Keenan, M. (2003): Two and a Half Cycles of Foresight in the UK, Technikfolgenabschätzung – Theorie und Praxis, Vol. 12, No. 2 (June), pp. 41–49

Petermann, Th.; Coenen, R. (eds.) (1999): Technikfolgen-Abschätzung in Deutschland – Bilanz und Perspektiven. Frankfurt u. a.: Campus

Porter, A.L. (2005): Foresight in Perspective, Prospecta Peru, 2005, Lima, (September)

Porter, A.L. (forthcoming): Technology Foresight: Types and Methods, International Journal of Technology Management, forthcoming

Porter, A.L., Ashton, W.B. (2006) United States Case Study, in L. Georghiou, J. Harper, M. Keenan, and I. Miles (Eds.) International Handbook on Foresight and Science Policy: Theory and Practice, Edward Elgar, Arnold, London

Porter, A.L., Roper, A.T., Mason, T.W., Rossini, F.A., and Banks, J. (1991): Forecasting and Management of Technology, Wiley, New York

Porter, A.L., Yglesias, E., Kongthon, A., Courseault, C., and Newman, N.C. (2004): Get What You Need from Technology Information Products, Research-Technology Management, Vol. 47, No. 6, pp. 16–19

Rader M. (2001) Monitoring of Technology Assessment Activities, report by the European Science and Technology Observatory Network

Rader, M., Boehle, K. (2003): First Findings from FISTERA on Foresight, Technikfolgenabschätzung – Theorie und Praxis, Vol. 12, No. 2 (June), pp. 75–78

Rader, M., Böhle, K., Hoffmann, B., Orwat, C., Riehm, U. (2003): Report on Findings on IST from Eight Selected National Foresight Exercises, (April), http://www.itas.fzk.de/eng/projects/fistera/wp1reps/d1.1-0304.pdf

Robinson, D.K.R., and Propp, T. (2006): Multi-Path Mapping as Strategic Intelligence for Reflexive Alighment in Emerging S&T, FTA 2006 Seminar (Future-Oriented Technology Analysis), Institute for Prospective Technological Studies, Seville, 28–29 September

Salmenkaita, J-P. and Salo, A. (2004): Emergent Foresight Processes: Industrial Activities in Wireless Communications, Technological Forecasting & Social Change, Vol. 71, No. 9, pp. 897–912

Staton, M. (2006): Monstrous Foresight, FTA 2006 Seminar (Future-Oriented Technology Analysis), Institute for Prospective Technological Studies, Seville, 28–29 September

Steyaert, S. Eggermont, S., Vandebosch, H. (2006): Towards the Desired Future of the Elderly and ICT: Policy Reccomendations Based On a Dialogue with Senior Citizens. FTA 2006 Seminar (Future-Oriented Technology Analysis), Institute for Prospective Technological Studies, Seville, 28–29 September

Tubke A., Ducatel K., Gavigan J., Moncada P. (Eds.) 2001, Strategic Policy Intelligence: Current Trends, the State of Play and Perspectives, IPTS, Seville

Von Schomberg, R., Guimarães Pereir, Â., Funtowicz, S. (2005): Deliberating Foresight-Knowledge for Policy and Foresight – Knowledge Assessment. European Commission Working paper. Available at: http://www.cordis.lu/foresight/working.htm

Wagner, C.S., and Popper, S.W. (2003): Identifying critical technologies in the United States: a review of the federal effort, Journal of Forecasting, 22 (2/3), 113–128

Chapter 4
Revisiting Foresight Rationales: What Lessons from the Social Sciences and Humanities?

R. Barré and M. Keenan

4.1 Introduction

Foresight is now a well-established tool used by policy makers, strategists, and managers. It has been widely applied at the national level by science ministries and research funding agencies for developing shared long-term visions, for setting research priorities, and for strengthening interactions within research and innovation systems. It is being increasingly utilised in regions to formulate regional science and innovation policies. It is also used in organisations – both public and private – for scanning future threats and opportunities, and for formulating and future-proofing long-term strategies.

A lot of claims have been made as to the benefits of using foresight in support of policy making and decision making more generally. These often point to things like the identification of spending priorities for government agencies to follow, though 'softer' benefits have been increasingly mentioned, such as network building, changing mindsets, building trust among actors, and developing better preparedness for change.[1] Many of these expectations around the benefits of foresight, both tangible and intangible, have been increasingly adopted and articulated by the managers and sponsors of foresight in the formulation of the objectives set out for their exercises (see Box 4.1).[2]

In spite of this expansion in expectations, it would be fair to say that there is little understanding of the mechanisms at play in generating these benefits. Their expectance is more a leap of faith. Moreover, as things stand, not only is there an absence of a better understanding of the dynamics of foresight, there is also little systematic evidence to suggest that many of the claimed-for benefits regularly accumulate from foresight activities. Several reasons for this lack of evidence have been cited:

[1] In fact, many of these benefits were already highlighted by Irvine and Martin (1984) more than twenty years ago, but it has taken considerable time for them to be widely adopted as formal rationales for using foresight.

[2] This list of common uses of foresight is based upon stated objectives and impacts recorded for a selection of foresight exercises mapped by the European Foresight Monitoring Network (EFMN) – see http://www.efmn.eu

C. Cagnin et al. (eds.) *Future-Oriented Technology Analysis*,
doi:10.1007/978-3-540-68811-2,

Box 4.1 Some common uses of foresight (Keenan and Popper 2007)

Informing decision-making processes

- ✓ Formulate funding and investment priorities for public policies
- ✓ Evaluate existing strategies against potential futures, and devise future-proof strategies
- ✓ Develop reference material for policy-makers and other actors to use, broadening the knowledge base around which decisions are made, thereby resulting in better informed public policies or organisational strategies
- ✓ Provide anticipatory strategic intelligence to innovation system actors
- ✓ Detect and analyse weak signals to 'foresee' likely future changes and to gain insights into complex interactions and emerging drivers of change
- ✓ Identify new S&T, business, societal, policy and political opportunities
- ✓ Increase awareness of possible risks, and hence the basis for more effective contingency planning, and the design and development of appropriate forms of resilience

Mobilising and structuring actor networks

- ✓ Improve implementation by enabling buy-in to decision-making processes
- ✓ Increase understanding and build trust between participants, thereby contributing to the building of shared agendas
- ✓ Develop widely shared visions of the future with which actors can identify and thereby better co-ordinate their activities, be they individuals or organisations
- ✓ Disrupt 'lock-in' thinking and challenge fixed mindsets
- ✓ Build hybrid networks and strengthen communities
- ✓ Aid communication, understanding and collaboration across boundaries, be they geographical, organisational or disciplinary in nature
- ✓ Deepen dialogue with society and improve governance

Creating new capabilities

- ✓ Enhance strategic capabilities of organisations by helping to develop a language and practice for thinking about the future – something that is often termed a 'foresight culture'
- ✓ Enhance the standing and image of organisations using foresight, showing them to be future-oriented and open, and attractive places for investment

- The objectives set for foresight are often wide-ranging and vague, making them problematic starting points for evaluation
- The intangible benefits that are said to accrue from foresight are difficult to assess in themselves

- The complexity of cause-effect relationships, which cannot be handled by the often overly simplistic models used when trying to understand and give meaning to foresight activities and their effects
- The systemic and distributed nature of foresight means that benefits are likely to be dispersed across a landscape of actors and systems making attempts to account for effects very difficult
- Many expected impacts of foresight take several years to materialise, and when they do, they are often dependent upon other factors, leading to attribution problems
- There are so many different methodologies and settings for foresight that it is difficult to arrive at standardised evaluation approaches
- The costs associated with evaluating foresight activities

The argument in this chapter is that whilst the objectives set for foresight are increasingly wide-ranging, the conceptualisations of these activities, as indicated by their 'intervention theories', are somewhat lagging behind the latest insights offered by the social sciences and humanities (SSH). This conceptualisation gap has led to a situation where foresight activities are insufficiently understood, or even misunderstood, making any assessment of their outcomes problematic.

At the moment, policy makers and analysts are still trying to better define the expected outputs and outcomes of foresight, based largely upon an empiricist approach of learning from case experiences. Whilst an empiricist (inductive) approach is worthy and most definitely necessary, the argument in this chapter is that it is insufficient on its own. We argue that to improve our understanding of foresight, we must turn to the SSH for a more deductive mode of enquiry.

4.2 The Promise of SSH?

We believe that, as a starting point, the following SSH disciplines could offer potentially useful lines of interpretative enquiry:

- From epistemology, the status of knowledge claims generated through the use of foresight methods; foresight knowledge production in the continuum from 'scientifically-certified' to 'socially-robust' to 'politically-relevant' knowledge.
- From political science, foresight as an instrument of deliberative democracy in its dialectic relationship with representative democracy; foresight as processes for defining the common good, in its relationship with private and public goods; the question of the legitimacy of foresight exercises in relation to the political system and the linkages between 'efficiency' and 'legitimacy'; the role of foresight in the evolving governance structure of knowledge societies, which are also multi-level governance structures; and foresight and the 'consequences issue', highlighting the question of responsibility.
- From sociology, foresight and the building of collectives: foresight as a process of co-production of communities of stakeholders and of the co-production of

regulation mechanisms; foresight as a mechanism to be understood in the context of the 'ecology of promises' of science and technology, that is the complex interplay between the 'goods' of S&T expressed by scientists to secure budgets and the 'bads' expressed by specific stakeholders, such expressions being the process which create the stakeholders ('concerned groups'); foresight as an instrument of transaction among social groups and stakeholders, allowing for the expression of different rationalities; and foresight and the recognition of the variety of the modes of knowledge production, including the production of 'socially robust knowledge'.

- From economics, foresight as a coordination mechanism of the agents in a knowledge economy; foresight analysed with the concepts of knowledge economics, including characterisation of the flows of knowledge in foresight exercises; foresight analysed with the concepts of evolutionary economics, for example, the role of foresight in the shaping of behavioural and adaptive routines of agents.
- From management and organisation science, foresight as a collective learning mechanism through the sequential interplay between codified and tacit knowledge; development of absorptive capacity and readiness to utilise foresight processes and outputs; the organisation and location of foresight operations in relation to issues of strategy and governance; relation of foresight to the development of anticipatory, inclusive and adaptive capacity.

This chapter seeks to make a contribution to better understanding foresight using concepts drawn from a broad range of SSH traditions. It is limited to consideration of just four theoretical approaches that could inform foresight planning, the expectations around foresight impacts, and, ultimately, the evaluation of foresight. The four focused upon here are as follows: (a) epistemology and the sociology of science; (b) organisation studies; (c) economics of knowledge; and (d) evaluation-utilisation studies. More will be said on this choice below.

4.3 Conceptualising Foresight

At a general level, the purpose of foresight is to allow for a better accounting of the long term in decision-making processes and to allow for collective processes of identification and debate of alternative strategies, particularly in public policy-making regarding research and innovation. The ambition of foresight is, in other words, to improve the two way linkage between knowledge and the building of a "common world". The key notions related to foresight are policy making, public participation, learning, alternatives, complex socio-technical systems, and science-society relationships. Foresight is the set of activities dealing with statements about long term future dynamics, either to produce such statements or to perform criticism of existing ones.

Thus, a foresight operation has two components: the production of conjectures and their articulation in socio-political debates. More precisely, foresight consists of (a) the elaboration, based on rational methods, of conjectures (hypothesis, statements) about the evolution and future states of the system considered; and (b) the structured organisation of debates about them and their elaboration (the foresight forum). This means foresight requires that the conditions of production of the conjectures are explicit and in such a way that they are debateable (Mermet 2005). Since the methods must be rational and the conjectures must be debated in a structured way, it follows that foresight operations must be conceived:

1. *From a large and open repertoire of methods and modes of interventions.* Foresight operations can be very diverse, since they can be characterised by the different relative importance given to the production of conjectures or to the structuration of the debates.[3] In addition, a wide range of possibilities in terms of methods exist, from the most standardised, to the most ad-hoc. The important aspect here is that a foresight operation can be described and characterised along its fundamental dimensions, i.e. the relative importance of conjecture production and debates, and the modalities of intervention related to each.
2. *With reference to normative principles ensuring such rationality and debate.* These principles are of two kinds: the scientificity principle and the democratic principle. The scientificity principle states that attention must be given to the epistemological status of the conjectures that are built, in order to define the relevant rules and criteria for debate and criticism. The democratic principle states the need for equal respect (status) among all the actors in the operation, bearing in mind the problems of asymmetries of power and their implications for the rules and criteria of the process of debate.

Taking these points into account, foresight can be considered as a producer of conjectures on the future, as deliberative process, as a learning tool, and as an influence on strategy formulation and follow-up action. This understanding of foresight suggests that many areas of SSH could offer useful lines of enquiry for a better understanding of foresight dynamics. For example, we propose (a) that exploration of the conjectures produced can be improved through reference to epistemology and the sociology of science; (b) that learning effects can be considered using concepts borrowed from organisation studies; (c) that the coordination of actor strategies can be analysed using an economics of knowledge framework; and (d) that the influence of foresight (for example, on public policy and actor strategies) can be investigated with reference to utilisation studies around evaluation and other sources of (expert) intelligence. We will therefore analyse foresight processes as seen from the point of view of these four approaches.

(a) *The quality of foresight in terms of statements production and debate: an understanding through the epistemology and sociology of science framework*

Foresight does not produce scientific knowledge since the statements about the future (conjectures and scenarios) that are generated are not falsifiable – in the sense that there

[3] This is close to the two dimensions defined for building a typology of foresight, namely the level of analytical elaboration and the level of participation (Barré 2001a).

is no way to design a procedure that could prove them wrong. One could object that this is the situation for any discipline where modelling is a central tool. Foresight is a special case, however, since it deals with large and complex socio-economic and technical systems. These prevent any modelling that can be called scientific, since the relevant parameters are far too numerous, intrinsically heterogeneous, and often non-quantifiable. Another way to put this is to recall the obvious fundamental and irreducible uncertainty of the long term future of systems having a social component.

Foresight either uses very partial and simplified accounting-type models to check coherence (ad-hoc modelling), or borrows existing models from a discipline (economics, demography, hydrology, etc.). However, these tend to address only a small fraction of the systems foresight usually deals with, and are at their limits of significance when simulations are made in the longer term and not 'all things being equal', which is usually the case with foresight. Nevertheless, these uses of models can be very useful and are often important in foresight processes, though it does not follow that the conjectures and scenarios produced are scientific results, nor that they can be labelled as forecasts, nor that the ad-hoc modelling taking place in many foresight exercises can be considered to generate scientific models.

In this sense, foresight is not based on an epistemology of predictability, nor on an epistemology of modelling; we can just say (and this is crucial) that it is based on a rational approach which follows the logic of modelling, which enables us to build statements about the future that are coherent and based on scientific knowledge from a variety of disciplines, allowing for the possibility to discuss the rigor of the process, its analytic sources, its assumptions, and the causal relationships considered. In other words, conjectures about the long term are statements based on scientific knowledge, but 'transgressing' it by applying it beyond the limits of its validity, the gap being filled with representations of the world and specific values (hence the notion of alternative scenarios).

This situation, as the sociology of science shows, appears to be a particular case of a discourse surrounding scientific expertise. In scientific expertise, the sources of uncertainty are both the extension of the topic beyond the discipline of the expert and the scientific controversies that appear usually at the frontier of science. In foresight, the uncertainty lies in the impossibility to know the future. It therefore follows that the conditions of its debate can be the same as a debate in an expertise process context.

The concepts of socio-technical networks and 'hybrid forum' can be used to describe the interactions and working of the heterogeneous sets of actors in a foresight context: the foresight actors can be seen as participating in an open knowledge and innovation producing process. This leads to the idea that foresight goes through the same sequences, namely the 'translation' sequence (Callon et al. 2001):

- The reduction of the 'macrocosm' (the real world) to the 'microcosm' (what can be handled in a laboratory for experiments)
- The work of a specialised and reduced collective, working on the microcosm, imagining and exploring simplified objects, based on a high concentration of instruments and competencies
- The return, always risky, towards the real world of the macrocosm: will the knowledge produced in the confined space of the laboratory survive?

In the case of foresight, the only difference is that, as we have seen, the second stage (laboratory experimental work) is of a different nature, dealing with exploration of hypotheses without procedures to close the spectrum of alternatives. The third stage provides an analogy of status between the statements on the future resulting from a foresight exercise and the 'socially robust knowledge' produced in the 'Mode 2' sequence (Gibbons et al. 1996).

(b) *Foresight as a collective learning process: an understanding through an organisation sciences (management sciences) framework*

Individuals and groups adapt their behaviour to the representation of the situation they face, which leads them to mobilise cognitive capabilities. In this sense, organisations are 'systems of interpretation', filtering and analysing their environment through collective representations. Behaviour can change through learning, though here there are two kinds of learning:

- Single loop learning amounts to direct adjustment, keeping existing references: behaviour changes but not its underlying values, allowing for evolution in a given framework.
- Double loop learning which results in behavioural change based on cognitive evolution. This leads to new interpretative schemes: the understanding and interpretation of the situation is changed and the underlying values are restructured, raising the "why?" question, thus addressing strategic aspects.

Such cognitive changes are the basis of collective learning, which requires reflexivity and 'distanciation' regarding one's prevailing representations: in other words, a global and systemic vision.

Tacit knowledge includes mental models based on personal experiences, which help individuals to perceive and give meaning to their environment. Such models structure the representations of reality of individuals, as well as its visions of the future, and are collectively shared in 'organisational cultures'. The dynamics of such knowledge is based on the interaction between codified and tacit knowledge.

Foresight is essentially a process by which professional communities interact and exchange both codified (formalised) and tacit (personal, embodied) knowledge in the sequence where knowledge is transformed from the (initial) tacit state into codified knowledge (through organised interactions among participants) and, after proper "treatments", which is feasible with such codified knowledge, transformed back into tacit knowledge through appropriation by the participants.[4] Thus, foresight is, strictly speaking, a double loop learning procedure: the collective search for 'drivers' of the system, the debates on the possible evolution of such drivers, the conception of scenarios or visions of possible future states of the system – all of these are typically collective interactions involving codified and tacit knowledge. In fact, all the methods used in foresight can be described in this way.

---

[4] Interestingly this process has similarities with the macrocosm – microcosm – macrocosm process of the production of scientific knowledge presented above.

(c) *Foresight as a process for the coordination of agents' strategy: an understanding through an economics of knowledge framework*

The dynamics of the new dominant sciences (NBIC – nano-bio-information-cognitive) (Larédo 2003) are characterised by a proliferation of research trajectories, coinciding with a decentralisation of initiatives and the multiplication of the institutions and laboratories in competition. The branching of relevant networks, the proper alliances, and the reactivity in front of opportunities or new information on the promising lines of research, become major factors of excellence and competitiveness.

In addition, the extension of interdisciplinary research and project financing from a variety of sources lead research institutions and laboratories to redefine their modes of working, their partnerships and their positioning. They have to acknowledge the moving and open frontiers between market and State, science and non-science, production and application of knowledge (Barré 2001b), and the distributed character of the processes of knowledge and innovation production. To add to this, they must also acknowledge the crisis of the modes of representation and action of the State.

Facing such challenges, the very bases of public action are being redefined. Thus, the classical justification of public intervention in terms of market failures[5] is being completed by a justification in terms of 'system failure', i.e. linked to the dynamics of the innovation system as such (Smits 2001). The new public action rationale states that crucial points include the management of interfaces among activities, the development of experimentation and exchange platforms, the circulation of talent, and so on. But there are even more important implications, since the performance of such a system is driven by the notion of 'distributed intelligence'. It implies that the strategic capabilities of the distributed agents is a central determinant of the overall performance, hence the need of a function allowing the emergence of spaces of expression and sharing of anticipations and coordination, recognised by all. This is precisely the function of foresight operations (Barré 2002).

The new terms of reference for public action are the following: to contribute to interactions that allow actors (firms, laboratories, non-profit organisations, and so on) to produce the proper information and visions necessary for developing their strategic intelligence to participate in the processes of knowledge and innovation production (Smits and Kuhlmann 2004). In other words, foresight operations represent a major function in the working of a knowledge economy: dealing with an immaterial resource, there must be mechanisms allowing for the revelation of the functional equivalents of supply and demand, quantities, and prices.

(d) *Foresight as Strategic Policy Intelligence (SPI): an understanding through conceptualisations of the use of evaluation and technology assessment studies*

Looking at the literature on the impacts and use of evaluation studies, a growing acknowledgement of the complexities at play is apparent. According to Weiss (1998),

[5]The social returns from research and innovation are higher than their private returns, hence the rationale for government intervention.

early studies on evaluation use carried out empirical studies to "identify the correlates of use, the characteristics of potential users, and communication strategies that were associated with greater use of results". This has some resonance with today's efforts at better understanding the use of foresight results. But, according to Weiss, evaluators came "to a growing realisation of how complicated the phenomenon of use is and how different situations and agencies can be from one another" (ibid.). Accordingly, the "old kinds of studies that tried to isolate the few keys to utilisation have largely gone out of style", to be replaced by an enhanced understanding of the use of evaluation through the "inheritance of new perspectives from other fields" (ibid.). In other words, through reference to SSH, evaluators have found new ways of thinking about utilisation, although as Weiss admits, "we have not solved the problem [of utilisation] but we are thinking about it in more interesting ways" (ibid.).

Conceptualisations of the meaning of the term "use" have also expanded. In the early days, they referred to an instrumental notion of use, with the information outputs of evaluation seen as information inputs to policy processes. This reflected a rational world-view of the policy process whereby evaluation was supposed to create new knowledge that would be absorbed by policy makers with the aim of improving policies and programmes. But empirical evidence that demonstrated this mechanism was hard to come by – if anything, it seemed as though the results of evaluations were largely neglected by policy makers.

However, by the late 1970s, it was already understood that policy processes rarely worked in rational ways and alternative models of negotiation and bargaining were proposed. In such settings, the utilisation of evaluation was re-conceptualised. Whilst instrumental use was still admitted, other uses were elucidated, including conceptual use, persuasive use, symbolic use, and unintended use. Viewed in this way, evaluation has real consequences, challenging old ideas, providing new perspectives, and helping to re-order the policy agenda. According to Weiss (1999), this kind of 'enlightenment' is difficult to see, but it is there nevertheless. It has been shown that "many channels bring evaluation results to the attention of policy makers, and they listen not only because they want direction, but also to justify policies, to show their knowledge and modernity, and as a counterweight to other information" (ibid.).

In the realm of foresight, notions of instrumental use still predominate, but this narrow conceptualisation should surely be expanded. Borrowing from evaluation studies, we might therefore conceptualise foresight use as follows:

- Instrumental use, where the outputs of foresight are used directly to inform decision making processes. This is the most common concern of sponsors of foresight, who want to see impacts from the foresight exercise they have supported. And concern for impacts is more often than not confined to consideration of the instrumental use of outputs, especially recommendations.
- Conceptual use, where the models and scenarios introduced in a foresight exercise change the perceptions and understanding of the issues covered, providing new ideas and insights. When conditions allow, this new conceptual understanding can be deployed in instrumental ways.

- Persuasive use, where foresight is used to mobilise support for a position that people come to hold about the changes they consider to be needed. For example, stakeholders and experts often agree upon the problems and challenges around an issue and the solutions needed to address them. They use foresight to legitimise their position and to gain wider circles of adherents through the building of more extended advocacy coalitions.
- Unintended use by non-target audiences, where the findings of a foresight activity might be drawn upon by foresight activities at other levels (for example, where companies or sub-national regions might use the results of national foresight exercises in their own foresight and strategic planning processes) or included in a meta-analysis of such studies (for example, the EC often elicits such studies looking across the results and processes of foresight activities carried out in the Member States).

We might also consider what it is that is actually used. Some possibilities include the following:

- Foresight processes generate findings, and this is usually the place people look when they want to evaluate implementation. More often than not, the focus is upon implementation of recommendations, with the evaluator 'counting up' the number of recommendations apparently taken up. However, foresight also generates a much broader array of findings, many of which are published as standalone reports that can be used in a variety of ways.
- The ideas and insights generated through the foresight process can lead to what Weiss (1980) has termed 'enlightenment' among participants and users. The extensive array of documented findings generated by foresight are often used in this way, whilst the dialogue and interaction associated with foresight processes may also lead to the creation and sharing of ideas and insights.
- The act of performance of foresight can, in itself, be used by sponsors (e.g. public agencies, companies, sub-national regions, etc.) to demonstrate their forward-looking and democratic (through promotion of participatory processes) credentials. Whilst this symbolic use of foresight might be viewed cynically, it is a fact of life and does not mean that the results of such activities are ignored.
- Returning to process benefits, those who participate in foresight often begin to think more expansively and more deeply about the issues being covered. Besides reflecting on current assumptions and thinking more critically, they may also adopt a more forward-looking and anticipatory outlook. Moreover, they do this with other actors, which can lead to alignment and commitment to shared action.

Building upon the work of Weiss and others, Henry and Mark (2003) have tried to set out a more systematic approach to examining the uses of evaluation studies. In fact, they prefer to replace the notion of 'use' with 'influence', which they say is more wide-ranging. Drawing upon areas of SSH that deal with change processes, including research in psychology, political science, and organizational behaviour, they have listed several specific types of change processes or outcomes that could be triggered by an evaluation. These have been ordered into three levels:

- *Individual level.* Refers to situations where evaluation processes or findings directly cause some change in the thoughts or actions of one or more individuals. Translating this to foresight, the process of participating in a workshop, a Delphi, a consensus conference, and so on, is believed to change people's beliefs and opinions and to mobilise them to act. Reading the findings of foresight is also supposed to have similar effects, although it is widely believed that process participation is better at achieving mobilisation.
- *Interpersonal level.* Refers to a change brought about in interactions between individuals. In a foresight context, this could refer to the personal interactions that occur in the 'hybrid fora' built into foresight, such as workshops, panel meetings, consensus conferences, and so on.
- *Collective level.* Refers to the direct or indirect influence of evaluation on the decisions and practices of organisations. Within the realm of foresight, this could refer to institutionalised policy learning or to the setting of spending priorities in light of foresight findings. In addition, given the importance attached to inter-organisational networking in foresight (particularly in foresight studies), this level of analysis could also cover the aggregation processes associated with collective action around shared visions.

4.4 Putting Theory into Practice?

The utilisation of SSH suggests that efforts towards better conceptualisations will largely need to be researcher/practitioner led. A first stage in this process will be to enquire as to the current intervention theories in use – not only among sponsors and policy makers, but also among researchers and practitioners. The following sorts of questions might be asked: how sophisticated are current intervention theories and what sorts of things do they cover? What models or concepts do they draw upon, for example, from innovation studies or other SSH traditions? Are these models and concepts appropriate? Indeed, are there other SSH models and concepts that could be drawn upon? We suspect that such interrogation of intervention theories used in many contemporary foresight exercises will show them to be rudimentary and seldom reflecting latest thinking in the SSH. It may be that some basic logic is articulated but this is unlikely to go so far as to explicate the underlying mechanisms at play.

The next step – which will be an ongoing process for some years – will be to try to better understand and model foresight using ideas and concepts drawn from a broad range of SSH traditions. On the basis of this understanding and conceptualisation, not only can better evaluation indicators be devised, but foresight practices themselves can be better designed for particular purposes. The latter is important, since foresight can be applied to a diversity of challenges in a wide variety of settings, and knowing which approaches are likely to work best offers the potential to improve practice.

References

Barré R. (2001a) Synthesis on technology foresight in Tubke A., Ducatel K., Gavigan J., Moncada P., (eds.) Strategic Policy Intelligence: Current Trends, the State of Play and Perspectives, JRC-ESTO, EUR 20137 EN, 2001, pp. 51–64

Barré, R., (2001b) The Agora model of innovation systems: S&T indicators for a democratic knowledge society, Research Evaluation, vol. 10(1), pp. 13–18

Barré, R. (2002), S&T policy-making for the future: new rationales, new design tools in S&T policies in Europe: new challenges and new responses, STRATA Consolidating Workshop, Brussels 22–23 April 2002, Report EUR 20440, pp. 86–134

Callon, M., Lascoumes, P. Barthe, Y., (2001) Agir dans un monde incertain – essai sur la démocratie technique, Seuil, Paris

Gibbons, M., Limoges C., Nowotny H., Schwartzman S., Scott P., Trow M. (1996) The New Production of Knowledge, Pinter Publishers, London

Henry, G and Mark, M (2003) Beyond use: understanding evaluation's influence on attitudes and actions, American Journal of Evaluation, vol. 23(3), pp. 293–314

Irvine J., Martin B.R. (1984) Foresight in Science, Picking the Winners, London, Printer Publishers

Keenan M., Popper R. (eds.) (2007) Guide to Research Infrastructures Foresight, Brussels, EC

Larédo, P., (2003) Six major challenges facing public intervention in higher education, science, technology and innovation, Science and Public Policy, vol. 30(1), pp. 4–12

Mermet, L (2005) Etudier des ecologies futures, PIE-Peter Lang, Brussels

Smits, R., (2001) The new role of strategic intelligence, in Strategic policy intelligence: current trends, the state of play and perspectives in Tubke A., Ducatel K., Gavigan J., Moncada P., (eds.) Strategic Policy Intelligence: Current Trends, the State of Play and Perspectives, JRC-ESTO, EUR 20137 EN, 2001

Smits, R., Kuhlmann, S., (2004) The rise of systemic instruments in innovation policy, International Journal of Foresight and Innovation Policy, vol. 1(1–2), pp. 4–32

Weiss, C. H. (1980) Knowledge creep and decision accretion, Knowledge: Creation, Diffusion, Utilization, Journal, vol. 3, pp. 381–404

Weiss C (1998), Have we learnt anything new about the use of evaluation?, American Journal of Evaluation, Vol. 19(1), pp. 21–33

Weiss, C. H. (1999) The interface between evaluation and public policy, Evaluation, vol. 5(4), pp. 468–486

Chapter 5
Monstrous Foresight

M. Staton

5.1 There's No Future in Foresight

5.1.1 A Strange Thing About Foresight

A strange thing about foresight is that you rarely hear anything about the future. In the futures work I have looked at and participated in the idea of the future 'as such', as the field or medium of our enquiry is, it seems, never mentioned. The future, while being the thing we are concerned with above all else, the thing we are trying to 'think' about, 'debate' and 'shape' in the words of the current European Commission definition of foresight on the foresight unit's website[1] is never the topic of theory or enquiry: absolutely everything else is, but not that. In fact, it is actually very difficult to say how the future is thought or theorised in most exercises, not at all easy to identify what idea of the future we are working with. At best, in most cases, it seems that it is considered to be unproblematic: a common-or-garden space and or time; empty but not quite a vacuum; waiting to be filled for good or ill by us or by others if we don't get there first; something that will open up and close like an infinite concertina depending on how hard we push at the walls of it (although it is accepted that things get less clear the more we push); shorter or longer depending on how we chop it up; something that while vaguely way 'out there' leads right up to our feet, through us contemporary subjects, back to the past. But mostly there is silence: the future is just a kind of unformed assumption at the middle of the discourse.

5.1.2 A Weakly Defined Concept of the Future Is a Problem for Foresight

Well, of course, we might say that foresight is not futurology and, of course, that's true. But, nonetheless, in all the definitions that we could collect up, the word future

[1] http://cordis.europa.eu/foresight/definition.htm. And here it is important to note that this paper is heavily Euro-centric – based mostly on work done at national and regional level in the EC.

C. Cagnin et al. (eds.) *Future-Oriented Technology Analysis*,
doi:10.1007/978-3-540-68811-2, © Springer-Verlag Berlin Heidelberg 2008

crops up somewhere as an organising principle, which sets up structures, builds hierarchies and conceptual geographies, calls in methodologies, puts them to work. So, foresight is about the future while having, as I just proposed, almost no concept of what that thing might be. This is a serious problem, for a number of reasons. Practically, for example, it is a barrier in the way of foresight ever becoming something more than general strategy development tool which seems to be where its most likely destiny lies at the moment. Of course, we can't expect foresight to account for itself fully, to have clear frontiers and territories, no discourse has those, but if it can't take account of itself and its key concepts or account for its own non-coincidence then it is still discourse in its infancy. If it can't stand up for itself and speak in its own name about what its business is it will mean that what is unique to it is never fully explored and the radical potential foresight seems to me to offer will become ever more diluted. And when the winds of research governance fashion change, as they will, it will be hard to mount a defence and foresight will disappear into obscurity again.

But I think there is a much more important reason why the weakly theorised idea of the future in foresight is a key issue. The future is the site of conflicting and competing discourses and ideologies, it is the site of politics, regardless of the claim that is sometimes made for foresight that it is a neutral space for debate and consensus formation. Foresight is a place where governments can and will try to stabilise, naturalise their roles, institutions jostle for positions, where sectors or fields fight for resources and where no one is ever likely to envisage their own demise, however inevitable or necessary or helpful as anything other than a crisis to be prevented. Therefore, unless the future is an idea along with other ideas and theories and not simply projected as a thing 'out there', fundamentally the same kind of stuff as the present here, then the artificiality, the partiality, the political nature of the constructions and the positing and populations, the exclusions, the damage, the territories claimed and openings closed down will never be thrown into sharp relief and the ever present danger will be that we may mistake how things are for what is yet to come.

In what follows I will try to show that a weakly developed idea of the future has also allowed foresight, by default, to become dominated by a discourse that in some respects isn't fit for purpose and which will help, perhaps, to ensure that there is no future in foresight in the more negative colloquial sense. I will then go on to sketch out another way of looking at things.

5.2 There's No Future Because we Won't Let it in

But first, I'll move on now to look a bit more closely at how, it seems to me, we try to speak about and manage the future in current foresight practices and how these approaches and methods create and reinforce a certain quite limited vision of what the future might be and effectively proscribe what types of things will we produce.

5.2.1 The Dangers of Dialogue

Current approaches to foresight emphasise discussion, consensus building, network building, inclusiveness, interactivity, process benefits, dialogue – we are all familiar and comfortable with these approaches and methods. I will call this the 'dialogic' approach and I think, perhaps surprisingly as it is all about open communication and exchange which we know to be categorically 'good things', that it leads us directly to the cause of the problems and limitations of many of the foresight exercises we see currently – the main problem being that there simply isn't very much new in the outputs of this work.[2]

In brief, I would suggest that it is very difficult, perhaps impossible, that the future, the future in a strong sense – the future proper,[3] a future other than an extrapolation of the present – can ever emerge out of this dialogic process. Of course, the process of dialogue cannot simply be reduced to a case of the 'blind leading the blind' – it is more open to ideas and phenomena than that and can clearly change and improve what people know by active interchange of ideas and experiences. But, I propose that dialogue in foresight is, fundamentally, based so squarely on rules of exchange we already know and don't question, on protocols of

[2] See later in this paper (Sect. 2.4) where I refer to research presented in Keenan and Barré's chapter in this book which suggests that one problem often reported by participants is that not enough new ideas are generated in foresight exercises. The question 'what's new?', however, is a fairly complex one. A good 'relativistic' question to pose for the findings of any foresight exercise is: 'new to whom?'. Putting this question gives rise to a typology of newness benefits such as that laid out by, for example, by Rémi Barré (in Tübke et al. 2001). This typology shows that foresight exercises which emphasise networking and process benefits tend to produce low levels of original knowledge but introduce new audiences to some ideas that are 'new to them' while more focused work with smaller expert groups will produce more knowledge that is not only new to participants but might be original. And it might be argued that foresight is useful if it performs the former as well as the latter. However, I'd like to hold to a slightly harder line here and argue that perhaps the only distinctive thing that foresight offers is an explicit concern with the new as the unforeseeable – all other benefits can be got by other means probably quicker, in particular the process benefits of the wider networked exercises which seem generic to all kinds of group work in policy making at national and regional level. I think that the idea of the new as the unforeseeable should be the 'benchmark' for evaluating the degree of foresight in projects and those with little or none of this element should be looked at very carefully to see if they could be done more effectively using other methods. So, we might ask of any exercise did it bring information already existing somewhere into a novel place to be encountered for the first time by a new community? If this synthetic role is its main characteristic then we might ask how could this be done quicker without bothering with foresight as it is a actually likely to be mostly a question of more efficient knowledge management and administration. Foresight, I argue here should concern itself only with what couldn't possibly be known, with questions other than distribution and dissemination of knowledge and, perhaps, with the new social configurations that might be got only through this focus on the new if process benefits are also considered to be part of the objectives of the activity. This stronger sense of the 'new' is unpacked in greater detail in the later sections of this chapter.

[3] For the idea of the future proper as opposed to the future perfect which is always already contained in the past (see Bennington 1994).

communication we are comfortable with and with routines for understanding, categorising and prioritising events as they are and realistically might be that all futures are subsumed by it and the future proper, the future unconnected to the present, is never allowed to arrive, can't break its way in.

Dialogue in foresight, is, oddly for a process which places such an emphasis on all forms of communication, barely a form of communication at all[4] – and the greater the number of people involved in the process the more this is the case as just managing the process becomes so difficult – tending rather to be more like a structured collection, dissemination and exchange of chunks of pre-formed opinion or fact than communication in a stronger sense. This stronger meaning of communication – 'communication proper' we might say – only happens where there is undecidability, risk, failure, confusion, potential violence i.e., where there is the need to communicate to address something undecidable which can't be dealt with simply by exchange, dissemination or the application of existing rules. If we don't need to communicate in foresight in this stronger sense then we aren't encountering anything new and the future is, above all, about what is new.[5] So, let's be aware of the dangers lurking within the apparently rational, progressive, democratic dialogic consensus if we are genuinely interested in encountering the future.

5.2.2 A Dialogic Future Is Tied to Today

To speak about the future all dialogic foresight tends to do is extend its conversations forward a bit. In fact, we are increasingly being asked as a condition of doing foresight at all, to make our future conversations ever clearer and more precise and

[4] For a sketch of a scene that throws light on the difference between communication in a strong sense and communication 'without needing to think' (see Bennington 1994).

[5] It is not likely that foresight as it is currently done in more open, network building type of projects which are aimed at developing consensus and shared vision will be improved very much or helped to reach their stated objectives by insisting on this type of communication as it is not an objective of this type of work and will most likely leave participants feeling that the experience has been unsatisfactory. Foresight processes that are more focused on knowledge production with less emphasis on process benefits are more likely to encounter and be more comfortable with stronger (even irreconcilable) differences of opinion and vision and probably have to manage this 'communication in a stronger sense' and it is likely to play an important part in these activities already although it has not been isolated as an evaluation criterion as far as I am aware. However, rather than wondering to what extent the type of communication described in this text might be included in all foresight types a better question might be: are those actions where we don't find the type of communication and the idea of the future as unforeseeable associated with it really foresight rather than resource planning or management activities? As is fairly obvious, I sense that foresight has suffered what project managers call 'scope creep' and has come to be more or less anything carried out in its name. In the bundle of activities which is coming to be known as FTA the role of foresight (perhaps even part of a tighter definition of what it is) might be to explore 'differends' (i.e., 'a case of conflict between (at least) two parties, that cannot be equitably resovled for lack of a rule of judgement applicable to both arguments' (Lyotard 1988), develop receptivness to the impossible future, the unpresentable, etc. But it would only make sense to try to do this where it will lead onto something like new knowledge and only in probably quite specialised contexts and not where it will bring processes to a grinding halt.

to account for them and their value by linking our dialogues back to action that we can recommend for the present.[6] In practice, what we are being asked to do is neutralise the future as far as possible, by sublating its difference under the concept of economic competitiveness or social well being or some other organising, probably totalising principle. Therefore, to be on the safe side, foresight tends too often to produce statements about management and allocation of resources, the reconfiguration or some small redistribution of things we know, of things we know how to do or can learn how to do. Foresight outputs are mostly about 'rolling out', about retargeting, writing new headings and subheadings, and all the while strengthening the links between the here and now, as commonly, pragmatically understand by policy makers with the future, as commonly and pragmatically not really understood at all by policy makers.

Foresight extends its dialogues into the future often by performing nothing much more than a change of tense to enable us to begin to order space and time by exchange. Therefore, often foresight seems to need to do nothing very much more than conjugate the present into the conditional tense, or perhaps, into the subjunctive mood, or sometimes in the future perfect (in forms of future writing or in scenarios) and occasionally into the imperative which for policy makers is, perhaps, best of all. Any lack of clarity we put down to poor future 'conjugation' through weak methodologies which need to be perfected or we lay the blame on something

[6] How this process works is presented excellently in Rémi Barré's chapter in Tübke et al. 2001. During a first exploration and hypothesis building stage (extension) participants engage in interactive activities. Extension is followed by a selection-convergence and synthesis stage (concentration) and this sequence leads to 'a description of Foresight as a learning process from tacit to codified knowledge transformation cycles…'. This paper also goes on to describe how in the most intensive and self reflective processes the values and assumptions that led to the production of knowledge in the extension-convergence process are also the subject of discussion and analysis; the so called 'double loop' learning. 'Double loop' seems to hold great potential to help foresight process get rid of the worse excesses of prejudice and political bias and would seem to be basic requirement for all work in this field, an interpretation of interpretations, a 'genealogy' of assumptions and values which can help us understand our attempts to stake out and occupy the territory of the future. However, in the face of the idea of the future that I am trying to hold onto in this paper in the latter sections, it is no more useful than any other methodology that is based on reflection, the 'double loop' is just two closed circles linked: the philosophical machinery of reflection is vast and full of assumptions about knowledge and subjectivity etc. etc. and it has usually been deployed to subsume and reincorporate difference. But 'double loop' learning does take us into what might be a productive line of enquiry: if we assume that the future is not available through reflexive processes then for all practical purposes we might be better to spend time making organisations and individuals as sensitive as possible to uncertainties and frontiers of current knowledge where they abut the unknown. This *can* be done through forms of reflection if done rigorously and continuously enough to make institutions highly attuned to change and better at dealing with it. It might be argued, and in fact this author believes, that public resources would be better off aimed at producing agile, chameleon-like institutions receptive to and welcoming of change based on the working assumption that for practical purposes the future arrives continually in small, strange, unaddressed but mostly just about recognisable packages that we can do something with rather than launching large, often isolated and static foresights groping optimistically, but melancholically, in a direction which is believed, or more accurately, hoped to be forward.

vaguely to do with the fact that with the further we go 'out there' the less likely it is that we can do things well, the less well we can see things as they recede into the distance.

5.2.3 And it Won't Help us Get Beyond the 'Longish Tomorrow'

Foresight can be pretty good at speaking about things that might happen tomorrow and for some vaguely specified time that can still linked firmly back to the present. But things don't go at all smoothly when we move beyond this 'longish tomorrow'. And things don't go well for more fundamental reasons than our failure to use dialogic methods well enough. Past the 'tomorrow +' problems become acute because we are trying to use the tools of dialogue, which are about knowledge, sharing, accumulation, learning from others, to speak about something which we can't know anything about in these terms. There is a fairly important paradigm clash, perhaps it might be called a clash of 'language games', going on here that helps explain a lot of the dissatisfaction with the more foresightful bits of foresight. When we try to use dialogic tools to talk about something (the future) that is radically not available to this type of conversation the results are always going to be unsatisfactory, we are never quite going to get what we were hoping for[7] – and there is a sense of the unsatisfactory, of anxiety, of unfulfilled promises that always seems to hang over foresight (perhaps I am the only one who feels this) and I think that this might be part of its cause.

So, it is clear that the future is something we want to get our hands on, we suspect it holds great riches, and we know it is not entirely the same as the present. But our tools are developed to deal with the present and the future, there-

[7] I propose this sense of unease – the sense of things never quite living up to expectations - surrounding foresight is a structural characteristic of it when we set out to look for the future with dialogical tools rather than a result of its not achieving any particular set of objectives. We can expect good learning impacts and a greater sense of a successful cycle of exploration, learning and 'double learning' and implementation being completed when the activity concerns the better distribution of existing knowledge; when the 'new to whom' question is answered by saying that knowledge has been brought to a new group of people although it might not be new to all. So, regarding *technology* foresight for example, we could use what is called foresight to help set more coherent and integrated research agendas and improve policy making in the sense of managerial efficiency or of reaching specific targeted research objectives if the very best knowledge was marshalled and the right process was in place; this is a system in which theoretically things could be altered for the better. But if we are speaking about the future in a stronger sense and foresight in a stronger sense, then we'll be speaking initially about the 'new to all' and dialogic tools will ensure that we never grasp whatever we set out to get simply because we have no way of discovering what we don't already know if we depend on dialogic processes. This problem is as old as the Western philosophical tradition: Meno's paradox in Plato is '…it is impossible for a man to discover either what he knows or what he does not know. He could not seek what he knows, for since he knows it there is no need of the inquiry, nor what he does not know, for in that case he does not even know what he is to look for.' In foresight we are still struggling with this problem.

fore, is either made to look like what we already know or it escapes us completely; either way there is nothing new to say. Therefore, I suggest that it might be helpful to think along another line, take a different starting point, with other presuppositions and a different idea of the future to produce different types of outputs that can be argued for in a different way, and help towards rejuvenating the future in foresight.

5.2.4 S&T Governance Is Vital, but Foresight Should Find its Unique Role

But before moving on, I'd just like to pause a moment to make it clear that, of course, none of this is intended to suggest that setting objectives to allocate resources against is unnecessary, clearly it is vital and we need to improve all areas of research and innovation governance and management as a key aspect of the drive towards getting more out of the money that we invest in these areas; the case of better agenda setting is clear and I won't take any issue with it, it needs to be done one way or another. However, I would like to suggest that we don't actually need foresight to do what we are doing with it currently, on the whole, and that most of the things we achieve through foresight exercises could be done by other means, or in fact, are being done by those other means which currently it is expedient to rename 'foresight'.

Here we can refer to Chap. 4 put together by Rémi Barré and Michael Keenan, where they discuss some of the impacts of foresight work collected on the extremely useful EFMN (European Foresight Monitoring Network) website. They mention: better informed strategies; making the case for increased R&D investments; creation of new networks and clusters; collective learning through open exchange of experiences etc. etc. I have just scanned the list at random and while useful, I am sure very useful in fact, I don't think that foresight is necessarily the only way or even the best way of getting these impacts, or that foresight wouldn't be better used doing something else as well as or instead of this because as resources are quite thin on the ground I think that we should find something that foresight does uniquely. Strategy development tools and planning methods as used in big enterprises, sectors or clusters of firms would, most of the time, bring us the same benefits as foresight brings currently in most cases. I suspect, for example, that if we were to import the very best agenda setting tools, perhaps from industry, we would achieve what we currently achieve, but probably better, quicker, in a more standard and, therefore, comparable and interchangeable way, allowing us, for example, to learn across regions in Europe, rather than encouraging endlessly different and mostly unconnected initiatives.

More importantly, Barré and Keenan's work points out a number of fairly important problems with foresight are beginning to become clear. We are hearing of the 'generation of visions that are too vague for follow up action', 'a lack of vision that results in few surprises, which calls into question the added value of foresight', 'insufficient learning across space and time resulting in much re-inventing of the wheel' etc. These are serious problems for the future of foresight and have the potential to undermine its claims to being a value added S&T governance tool and they must be addressed urgently as foresight continues to grow to ensure that it continues to be relevant. Keenan and Barré also quote experts who suggest that these problems can probably be overcome by more research and better understanding of the dynamics of the process. However, I propose that this is not the case and that to expect these problems to be solved by further research is to misrecognise slightly what is happening. The difficulties that we are seeing with foresight are, in fact, characteristics of the way that it is being done now rather than problems that can be solved with more research. I suspect that more research into how to make the foresight we have work better will mean us banging our head up against the wall to try to solve problems that we are better off viewing as likely inevitable by products of a certain type of foresight exercise, foresight based, as I suggested above on dialogic or vaguely dialectical principles hemmed in by demands for clarity and precision.

5.3 Is There a Future for Foresight?

5.3.1 Beginning to Think Differently About the Future

So, if the future is not coming through because of the structures we are using to try to think it and the framework in which it is taking place, let's try to think the future a bit differently. In the work of the French writer Jacques Derrida we find some ideas about how we might look at the future differently, for example from an early text:

> The future can only be anticipated in the form of absolute danger. It is what breaks absolutely with constituted normality and can therefore only announce or present itself in the form of monstrosity. For this world to come and for what in it will have shaken the value of sign, speech and writing, for what here guides our future perfect, there is as yet no exergue. (Derrida 1976)

This idea of the future is quite different to the one we are used to in foresight and is, first and foremost, something that can't be grasped by simply extending familiar ways of thinking or reduced to a space for the future implementation of programmes or protocols. Here the future breaks absolutely from the present and we can't extrapolate from where we are to what might be to come. There is no way for us to decode this future, no step by step strategy to capture it, no authoritative resource to employ for interpretation, no experts to turn to for explanation of its likely unfolding. This is a future we can't know anything about for sure, we don't get it, it upsets our speech and writing, our dialogues, it is open and unpredictable, and it appears as a danger, as a 'monstrosity' because it emerges for the first time, because we are unprepared for it and because of the threat and the risk it presents to everything we consider to be normal. It is a thing we don't recognise, a creature that fits none of our existing typologies of which Derrida (1978) writes elsewhere 'the as yet unnameable which is proclaiming itself and which can do so...only under the species of the non-species, in the formless, mute, infant, and terrifying form of monstrosity'.

If the future can only announce itself as something we can't grasp, categorise, sublate under the terms of dialogue and relate confidently to the present then a break emerges between the present and the future which opens the possibility, in fact, calls for a different and stronger type of foresight than the one we are used to. If we accept that the future breaks absolutely with the present then thinking about the future demands that we find new ways of exploring and speaking about what is to come, beyond dialogue and exchange, that would help us in the task of becoming receptive to the unknown and undecidable. How do we speak about the future without yanking it continually back to what we already know?, how do we make statements about it, perhaps even judgments without a stable foundation for speaking and judging?, these are tricky questions, they are probably in fact, questions of ethics primarily, but they are also questions for foresight, I think, and we should start to find a way to address them.

5.3.2 Good Foresight Might Be, then, to Chase Blindly After Monsters

In the peculiar light of this version of the future we must think again about the purpose of foresight, how it understands itself, its objectives and modus operandi. We might go as far as to say that, paradoxically, the proper business of foresight is really that which can't be foreseen, that which is formless and uncertain for the simple reason that because if it is not concerned above all else with the unforeseeable, foresight simply becomes 'sight'. Perhaps, we might concede, foresight depends on a certain kind of blindness because if we can already see things it serves no purpose beyond dissemination, advertising, perhaps propaganda. And if foresight depends on the unknowable we, in turn, only need it, perhaps, can only really use it to find the things that escape our dialogues and consensus, things we can't administer with existing rules, that don't conform to our typologies, which might in fact destroy our administrating and typologising, the things we in fact really might need to try to know about, the risky stuff, the dangerous things, the monsters.

In my opinion this necessary blindness of foresight is a very good thing because using this tool that depends on the unforeseeable to explore the unknowable and undecidable future actually seems to provide a very good methodological fit and I think, therefore, foresight has great potential for addressing the future proper if it is set up in a slightly different way to most of the exercises we see currently. It would be a very simple, a deceptively simple thing, a search for questions to which we don't already know the answers, for ideas that are new, to learn to be receptive to things we can't subsume under familiar concepts. However, such a practice would be stubbornly resistant to any idea of calculable return on investment and, therefore, we'd be swimming against the tide.

5.3.3 That's All Very Well But...: How Would We Justify This Approach To Funders?

So, how on earth would we go about justifying actually setting out, as we would have to, to look for things that don't make any immediate sense, or in fact, potentially don't make any sense at all as on the frontier of the future in the stronger sense it is impossible to tell the law bringer from the charlatan – it is the policy maker's nightmare. Well, firstly, to step back a little, I am not suggesting here that all foresight needs suddenly to become more radical and less directly linked to policy concerns, of course not.[8] I think it would be a good policy and method mix if there were a range of different options in play, but including undirected, exploratory monster chasing that was not expected to be value for money in the short term and that is not directly linked to current concerns even though such a practice would go against the dominant doctrine on foresight.

A justification for this type of activity can be built by questioning the ideology of science, research and knowledge that underpins and is deeply entwined in current thinking about foresight as a policy tool and which probably gave impetus to the new growth in popularity of foresight over the last ten years or so. This vision of science might be caricatured as a prejudice that science must become more accountable, provide value for money, get closer to the economy, closer to the concerns of the populace in the 'new social contract' that we heard of a few years ago (for example, Martin 2003; Gibbons 1999), closer to the interests of business, that science must be about impact, outcomes. All this now seems obvious and perhaps it is the dominant discourse about science that we

[8] I suspect that foresight will sooner or later have to find a firmer footing if it is to evolve and survive changes of policy making style (at the moment it is very much at home in networking/learning/capacity building/clutering 'paradigm' of policy thinking) when it will have to demonstrate its unique value added distinct from the merging and blurring of types of activities under the general heading of FTA; I don't really hold with the idea sometimes expressed that the openness and flexibility of the family of activities under the FTA heading is a strength, it seems to me to be a sign of immaturity and of making a virtue of necessity in the absence of very clear knowledge of what results each of the parts are capable of reliably delivering. Foresight, I propose should be about the unknown, the currently impossible (whether this unknown is 'in the future' or very close to hand – foresight isn't, in fact, about time at all in a simple sense) and other types of work should deal with strategy, resource allocation and dealing with programming issues as they do this more effectively which means that I think we need to aim for more distinctness, more specificity and specialisation in the tools that we are using currently and the projects that we launch.

hear in conferences such as this one where I haven't heard it challenged in years, perhaps ever, in fact. To suggest that this version of science and research isn't self evidently true, here, for the sake of simplicity I'll bring in some ideas from another French writer J. F. Lyotard whose discussion in his book The Postmodern Condition deals directly with science and the increasing demands that it become ever more 'performative' and helps us towards some kind of justification or 'legitimation' of a different vision of science practice that might in turn help us develop arguments to speak on behalf of a different foresight practice.

> Postmodern science – by concerning itself with such things as undecidables, the limits of precise control, conflicts characterized by incomplete information, "fracta", catastrophes, and pragmatic paradoxes – is theorizing its own evolution as discontinuous, catastrophic, nonrectifiable, and paradoxical. It is changing the meaning of the word knowledge…it is producing not the known, but the unknown. And it suggests a model of legitimation that has nothing to do with maximized performance, but has as its basis difference understood as paralogy. (Lyotard 1984)

Paralogy (literally, beyond or against reason) being here, at its simplest, the search for instabilities, the production of the unknown, the creation of ideas, science legitimised by, or argued for by its ability to chase what vanishes from us, what resists, what is left over, what has been suppressed, its ability to escape the thought systems we have, our routines and protocols, promoted as the place of what might be rather than what is. This is not a justification for research that most policy makers are used to hearing these days, but it is a good one, nonetheless. Could this form a basis for arguing for excursions in monstrous foresight beyond the long-ish tomorrow? I think that it could, in certain niche situations, as a complement and counterpoint to the dominant modes.

5.4 What Might this Mean in Practice? Some Preliminary Thoughts

5.4.1 What Kind of Work Would it be?

But what kind of activity would foresight legitimised by paralogy and based on an idea of the future as radically other be like? It would have to be designed as a search for new ideas, that might shock, surprise, perhaps disturb, for things that might not even look like ideas, but detritus, residues, echoes, flotsam as it is very difficult to tell the difference between the helpful and foolish when something is new. Therefore, the demand that it should produce economic returns will have to be put to one side in the same way that we still reserve large sections of science activity to curiosity driven research. Foresight would tend, therefore, to become something more like the activity of research itself rather than a supplement to it or a tool for managing it. Like any research[9] it would have a high tolerance of outputs that might not link directly to current issues but which form a stock of possible knowledge, a common resource of questions that don't come tagged with answers, the basis for reflection, further research, perhaps catalysts, perhaps new guiding threads towards a future unemerged, perhaps epiphanies, perhaps rubbish. This foresight would be done on behalf of knowledge and the yet to come not in the name directly of institutions or governments or cultures. It would be legitimised, perhaps, by the extent of its paralogy, by the strangeness of its outputs, precisely by our ability not to do anything with them, their resistance to immediate understanding, the extent that they demand different rules of engagement, draw out communication in a strong sense, not because they are vague or poorly produced or because the process has been badly done, but because we don't know what they are, yet. It would not, in short, be a management and resource allocation methodology or strategy development under a different name.

5.4.2 *What Kind of Tools Might we Employ? Inhuman Foresight Machines?*

We need a machinery that is not dialogic to learn how to speak about and welcome the unformed future. We need to get the subject and subjectivity (and the greater

[9] Of course, even curiosity driven science is a particular type of a discourse and always governed by rules of law, ethics and other conventions which delimit and determine the boundaries of its fields and methods, the places where it is done and the people who do it; it not in any sense unregulated, even the nature of failure and the protocols of its toleration are proscribed as far as possible. And I think that foresight of the kind I am sketching here would benefit from the rigour of scientific method and the self-regulation of scientific communities to a very large extent as I think the objective should be to develop a symbiotic relationship with foresight becoming some form of scientific or technical exploration itself (it probably already exists in fact in various forms in scientific and technical discourses) rather than tending to take a distance from these subjects to speak (mostly quite conservatively) on behalf of 'the community', 'values' or more often 'competitiveness'. So, something like this form of foresight would need to be done professionally and be recognised and valued by participants and stopped immediately if not. But it is difficult to see how a discourse setting out to explore paralogy can be regulated in advance to ensure that foolish or dangerous ideas or people are excluded as all new ideas are neither true or untrue, ethical or unethical as they arrive; there is not metalanguage to authoritatively explain new things and foresight would have to be open and receptive to the risks. In this respect a 'new foresight' would be quite different to current foresight work which mostly sets out to resolve risk in advance of it emerging in the name of what is often a quite normative regulatory idea, e.g., 'security', 'well-being', the mysterious 'European social model' etc. Of course, we can prepare ourselves and design processes as well as possible but an indicator of good design in this context would be an acceptance of the unregulatability of risk. Contrary to most current foresight there could be no priority on making science more open to wider, perhaps even public debate and to help it gain its legitimacy from this, in fact, quite the opposite, it would tend to push already difficult disciplines out to their limits. But it seems to me that foresight is not an effective mechanism for doing 'public understanding of science' work although this is probably important.

subject and group subjectivity of dialogue) out of the system and we need a method to break the gravity pull of consensus if something new can emerge or we can decide to do new things. Here is Derrida again with a quotation that links back to the earlier one on the future and helps to set up the kind of non-dialogic machinery that we might use for foresight:

> Doubtless the subjectivity of a subject, already, never decides about anything; its self identity and its calculable permanence turn every decision into an accident that leaves the subject indifferent. A theory of the subject is incapable of accounting for the slightest decision...nothing ever happens to a subject, nothing worthy of the name 'event'. (Derrida 1997)

The main idea here is fairly simple: if they are decisive and new we don't make decisions or encounter events as fully self-present, self-contained subjects, the 'I' or 'me' or 'ego'. The reason for this is that however well informed and carefully prepared we might be, the instant of decision (if it is worth that name) can never be explained entirely by what led up to it, it is heterogeneous to it, nor can the arrival of the new, the future (if it is worth that name) be anticipated. Decisions are always a leap beyond knowledge and calculation and receptivity to events is an opening out towards something beyond the subject's control unless it is simply a predictable thing that we already know or part of a programme we are administering under known rules where decisions are not needed nor the new encountered. And foresight fundamentally partakes of the structure of decisions and receptiveness to events to the extent that it is about going beyond the present, entering the realm of risk and uncertainty, making statements about what might be to come, encountering new ideas. So, I'd like to maintain that the future and its unforeseeable events which is the core business of foresight arrive to that which is other than the rational, self-present subject within us. I would also like to suggest that we can say the same foresight group work process based on dialogue whose apparent main strength is its capacity to bring together autonomous, rational, listening, speaking, open-minded groups of individuals to reach flexible, imaginative, well grounded decisions but which has the paradoxical effect of excluding new events and ideas and reducing decision making to programme management. So, in order to try to do justice to the future and to find a way of speaking about it we need a mechanism that reduces as far as possible the influence of subjectivity and that can perform the function of the other within us, the place of decision which we depend on if we are to encounter any new events and ideas.

5.4.3 A Practical Mechanism?

What kind of practical machinery can we use to help us welcome an unknown future? Well, this needs a lot more research, but I would suggest we look again at the big Delphis that were run previously mostly at national level and which have very much fallen out of favour these days. In fact, I suspect Delphi feels so unfashionable right now partly as a result of the fact that it doesn't throw up enough useful performative information and that there is a certain kind of madness in some of the statements. It is some of this strangeness that I'd like to see back, and which I would like to see magnified many times.

I think there is great potential to be explored in the 'inhuman', machine-like aspects of Delphi methods, where we participate, contribute, shape and yet don't control and whose transformative power has the potential to cope with and manage, fuse, amplify, distort a very large number of varied inputs, perhaps greater in number and complexity than subjects could cope with even in well managed groups, and play back to us things beyond the limits of our knowledge where perhaps just through sheer numbers of new linkages and permutations of possibilities, the unknown may emerge from out of the known. I think that the iterative, anonymising, quasi-automatic machine-like nature of the Delphi process probably augmented by the use of ICTs[10] can break us out of the orbit of dialogue and consensus and has the potential to generate ideas that are unforeseeable, if we prime and feed it with the right inputs and it is and managed against a 'paralogic' performance specification. But clearly this is just the lightest of sketches of a possible idea and a topic that I think deserves some more research.[11]

5.5 Conclusion

I'll finish with another quotation from Derrida which, while speaking specifically about the arts and humanities is also relevant to science and technology.

---

[10] Here we might look in detail at the potential contained in the way ICTs are used in the Rand study *in* Lempert et al. 2003.

[11] The toolbox of foresight is weighty and there are many ways in which these tools can be deployed that might take researchers where this paper suggests it might be interesting to go. Horizon scanning, for example as it is currently being done in the UK national foresight group of activities appears to be the kind of very open and diverse process that might be able to make similar kinds of findings to the ones I am gesturing towards here. I haven't been close enough to this work to trace in any detail how far it covers the kind of activities outlined in this paper and it may well cover the ground very well; it certainly seems to be a fascinating group of initiatives, in particular in the way that the knowledge appears to be managed as pool from which unforeseen connections could emerge or the data mined and linkages forged and knowledge created beyond the control of the managers of the datasets or any of the individual contributors: here perhaps is a primordial stew from which new knowledge might spring. The other method that is often mentioned to me as the answer to the kind of problems that I am setting up in this paper is wild card analysis but I am much more sceptical about this than horizon scanning although I should add that I have no experience of using it nor, therefore, a very detailed understanding of what it is supposed to be able to do. But I would insist that developing receptivity to futures really does depend on finding new forms of action that are not based on dialogic processes and if wild card work involves nothing much more than challenging groups with 'surprises' for them to find ways of dealing with these uncertainties then it seems that all that does is to improve the power of existing group processes to subsume the future while leaving all other assumptions intact. However, exactly how the work I am outlining here might fit into the wide range of foresight approaches and methods and whether or not what I am trying to grope towards here is already habitually done by other means will have to be the subject of further research.

> Texts and discourses that provoke at the outset reactions of rejection, that are denounced precisely as anomalies or monstrosities are often texts that, before being in turn appropriated, assimilated, acculturated, transform the nature of the field of reception, transform the nature of the social and cultural experience, historical experience. All of history has shown that each time an event has been produced, for example in philosophy or in poetry, it took the form of the unacceptable, or even of the intolerable, of the incomprehensible, that is, of a certain monstrosity. (Derrida 1995)[12]

I think this paragraph captures very well the idea of why we need to think of a way of welcoming what he calls in an earlier section of the same interview the 'monstrous arrivant'. Of course, any engagement begins the process of domestication and assimilation of the absolutely odd, we can't avoid this. However, the transformative power of the future is to be found in the anomalous, the intolerable and the incomprehensible before it is appropriated and turned into material for dialogic exchange and this pre-dialogic ground is the proper domain for foresight. Therefore, we should explore ways to develop a practice that meets the future as far as possible on its own terms, and it is impossible to know what those terms are in advance, and helps us to engage as quickly as possible and as openly as possible with transforming events that are yet to come. And with this call for an anticipatory foresight we have arrived, after a small detour, firmly back to the mainstream and to one of the key concepts that defines the field and which we all share.

References

Bennington, G 1994, Legislations, Verso, London
Derrida, J 1976, Of Grammatology, trans. G C Spivak, Baltimore and London
Derrida, J 1978, Structure, Sign, and Play in the Discourse of the Human Sciences' in Writing and Difference, trans. A. Bass, Routledge, London
Derrida, J 1995, Points-Interviews 1974–1994, Stanford University Press, Stanford
Derrida, J 1997, Politics of Friendship, trans. G. Collins, Verso, London
Gibbons, M 1999, Science's New Social Contract with Society. Nature Supplement 402, C81–C82
Kuhn, TS 1962, The Structure of Scientific Revolutions, University of Chicago Press, Chicago

[12] There are a number of similarities here with Kuhn's idea of the scientific 'paradigm' and 'paradigm shift' (Kuhn 1962) which might help orientate readers more familiar with the philosophy and sociology of science. In particular perhaps the idea of 'revolutionary science' when anomalous results begin to emerge in an existing paradigm creating the conditions for the emergence of a new paradign which is 'incommensurable' with rival frameworks could be thought of as quite similar to the 'monstrosities' that Derrida mentions in this paragraph. These points of similarity might be the an excellent place to start the articulation of these two bodies of thought in more detail which is far beyond the scope of this chapter. However, I suspect a point of key difference between the two thinkers will be that for Derrida the unceasing experience of the undecidablity of the new and the future which accompanies the imperative that we continually decide and respond as part of everyday life (obviously we have no choice about this) amounts to something like an ethics of receptivity towards the future while Kuhn's work has less of this urgency about it.

Lempert, R 2003, Shaping the Next One Hundred Years: New Methods for Quantitative, Long-Term Policy Analysis, Rand Pardee Centre

Lyotard, J-F 1984, The Postmodern Condition: A Report on Knowledge, trans. G. Bennington and B. Massumi, University of Minnesota Press, Minneapolis

Lyotard, J 1988, The Differend: Phrases in Dispute, trans. Georges Van Den Abbeele, University of Minnesota Press, Minneapolis

Martin, BR 2003, The Changing Contract for Science and the Evolution of the University. Science and Innovation. A. Geuna, A.J. Salter, and W E. Steinmueller (Eds). Edward Elgar, Northampton, MA

Tubke A, Ducatel K, Gavigan J, Moncada P (Eds) 2001, Strategic Policy Intelligence: Current Trends, the State of Play and Perspectives, IPTS, Seville

Part II
FTA Applications

Chapter 6
Technology Foresight as Innovation Policy Instrument: Learning from Science and Technology Studies

P. Warnke and G. Heimeriks

6.1 Introduction

There is a variety of interpretative frameworks for giving meaning to FTA activities (see Barré and Keenan in this volume). In this contribution we would like to explore an interpretation of Foresight from the perspective of the interdisciplinary body of knowledge that has become known as STS – Science and Technology Studies (c.f. Jasanoff et al.1994). Drawing in particular on STS insights on the "social shaping of technology", we would like to investigate the possibility of Foresight to support policy makers in influencing innovation trajectories according to societal needs.

The study of technological developments is a complex issue. First of all, technologies are not given in nature, but man-made constructs; they are the products of cultural evolution. The various actors involved may use different definitions of technology. Furthermore, technologies are continuously evolving in a social context. With the further development of technologies, their definitions and relevant perspectives may also have to change. These definitions and perspectives, however, are basic to the discursive traditions studying technology and its relevant contexts. We distinguish several perspectives in the study of technology: STS dealing with social and economic co-evolution of technology, and science and technology (S&T) policy analysis and R&D management through foresight. We argue that the combination of these three perspectives challenges us to consider technological change as a complex and reflexive process. From this combined perspective, it is implied that taking a holistic view on the future by looking at societal and technological elements together and aligning or even integrating companies and users visions on the future is not at all an easy thing to do. To actually develop socio-technical future visions Foresight needs to look at societal development and technological possibilities with the same degree of openness and expertise. STS results may give some indications for Foresight practice aiming to adopt such a holistic approach.

In the first section of this chapter we will briefly outline how Foresight is interpreted as (1) systemic innovation policy instrument fostering innovation capability, (2) orienting innovation towards societal needs, (3) agenda setting process and (4) a provider of anticipatory intelligence as a base for decision making. Starting from the assumption that the ability to guide policy makers on the implications of technological

C. Cagnin et al. (eds.) *Future-Oriented Technology Analysis,*
doi:10.1007/978-3-540-68811-2,

innovation for the wider socio-economic framework would need to be based on a clear understanding of the interplay between technological and socio-economic change we will then turn to Science and Technology Studies that deal in particular with this issue. We will briefly summarize how STS scholars have characterised technological innovation as a co-evolutionary process between technology and society and how the use of technology is described as a relevant feature of the selection environment shaping the direction of technological trajectories. STS is characterised by an insistence that the 'black-box' of technology must be opened, to allow the socio-economic patterns embedded in both the content of technologies and the processes of innovation to be exposed and analysed (MacKenzie and Wajcman 1985; Bijker and Law 1992). SST studies show that technology does not develop according to an inner technical logic but is instead a social product, patterned by the conditions of its creation and use. We will focus on the issues of Contingency and constraint of variation, Role of expectations and visions, Importance of downstream phase of innovation, Importance of localisation and Insights on steering possibilities for technological trajectories.

In the final sections, we will then again turn to Foresight and ask what lessons can be derived from these insights for the possible role of Foresight in supporting policy makers in intervening into this co-evolution process. We will elaborate one the concepts of foresight as a process moderator, foresight as expectation management, provision of anticipatory intelligence and localisation through Foresight.

6.2 Foresight and its Role Within the Innovation Process

6.2.1 Focusing on Foresight: Some Definitions

When discussing Foresight we are referring to a systematic process of reflection and vision building on the future among a group of stakeholders. The characteristics distinguishing Foresight from future studies and other long term thinking approaches such as strategic planning are (cf. e.g. Havas 2005):

- Participatory – In a Foresight process the relevant stakeholder groups are actively participating. Foresight results are disseminated and debated within a wider audience.
- Action oriented – In Foresight there is always the attempt to link insights about the dynamics of change to today's decision making, e.g. by elaborating strategic options to reach certain objectives or development of robust strategies to prepare for different future developments.
- Open – Foresight does not aim to predict a predetermined future but explores how things might evolve in different ways.

Foresight processes may take place in any kind of organisation to orient long term strategy building or to foster future oriented attitudes. However, in this paper we are focusing on Foresight activities that are carried out in support of policy making.

Over the last 10–15 years Foresight has been used in support of policy-making not only for research and innovation policy but also in other policy fields. Policy oriented Foresight exercises have been addressing a wide range of different issues. A rough distinction can be made between exercises focusing on a geographical territory such as a region or country, those that focus on a certain socioeconomic domain such as an industrial sector or a policy field such as transport and those that deal with a certain problem such as disease prevention (thematic Foresight).

While in the early years Foresight was mainly aiming to assess technological developments in order to inform priority setting in research policy and therefore tended to focus rather narrowly on technological and scientific developments the majority of exercises today is taking into account a wide range of social and economic aspects related to technological developments (Georghiou 2001; Salo and Cuhls 2003). Accordingly, the term "Technology Foresight" has gradually been dropped in favour of simply Foresight. However, research and innovation policy as well as technology policy are still the main clients for Foresight which means that even if a broad socio-economic view is adopted the recommendations for action will often target technology policy. Secondly, with technology having become deeply entrenched into modern society, reasoning about the future within almost any policy domain will involve thinking about technological aspects. Therefore, with the impact of Foresight on innovation processes and technological trajectories we are addressing a core dimension of the field.

6.2.2 *Role of Foresight Within Innovation Processes*

Whereas early Technology Forecasting approaches were based on a rather linear understanding of the policy process on the one hand and the innovation process on the other, Foresight is nowadays conceptualised as one element in a continuous policy learning process that is contributing to a more reflexive mode of policy making (Weber 2006). At the same time innovation is no longer understood as a linear process where research spending automatically leads to innovation and application with certain impacts on society. Both these insights imply that giving recommendations for research funding priorities will not necessarily result in any innovation activity or effect changes in technological pathways (Salo and Cuhls 2003). Accordingly, the way Foresight is thought to impact on policy has become more subtle and complex. We would like to highlight four modes of policy support Foresight is expected to deliver:

1. Foresight as systemic innovation policy instrument fostering innovation capability
2. Foresight orienting innovation towards societal needs
3. Foresight as agenda setting process
4. Foresight as a provider of anticipatory intelligence as a base for decision making

1. Foresight as systemic innovation policy instrument

Policy researchers have for some time now been stressing the need for "systemic" innovation policy instruments to complement classical tools such as direct research subsidies or public procurement (Smits and Kuhlmann 2004). These instruments are meant to enhance the capability of innovation systems for self-organisation so they address innovation policy on a system level. Systemic instruments aim to provide platforms for learning and experimenting, facilitate the management of interfaces, foster new alignments of elements, and stimulate demand articulation, strategy and vision building.

The need for systemic instruments is driven by a number of structural changes in the socio-economic framework of innovation activities and in particular changes in the speed and modes of knowledge generation (c.f. Tubke et al. 2001). Based on empirical research on various national and sectoral systems of innovation the capability to innovate has increasingly been characterised as a system capability where the connectivity between various elements (such as universities, firms, research institutions, government bodies) is of the same importance as the quality of the elements themselves. This notion of "innovation systems" as a core concept in explaining innovation capability is forming the background for the approach of systemic instruments.

At the same time the nature of policy making is changing. Due to the increasing system complexity, the traditional linear model of policy making incorporating successive phases like conceptualisation, implementation, evaluation and then modification and new decision making is no longer adequate. Policy and strategy formation is becoming more and more a continuous learning process (Lundvall and Borrás 1998). These changes are complemented by the emerging new models of governance such as multi-level governance driven by political challenges such as the EU integration (Kuhlmann 2001).

Within this framework Foresight is positioned as a systemic innovation policy instrument. It is argued that by establishing linkages between actors and providing platforms for joint learning Foresight helps to improve the ability of the system to react to changes and thereby to initiate and keep up innovation processes. Already in 1999 Martin and Johnston proposed to place Foresight in the framework of the NIS (national innovation systems) approach and described the crucial function of Foresight for innovation as "wiring up the innovation system" through "strengthening the connections within the national innovation system … and the system as a whole can become more effective at learning and innovating". This is very well in line with the notion of systemic innovation policy instruments serving to foster interaction between various actors of innovation (Tubke et al. 2001) such and users and producers of innovation, actors from scientific and technological realms or different disciplines and professional background as well as policy makers from various related policy fields. Foresight, so it is argued, fosters the flow of knowledge among all these actors and increases connectivity and coordination (Webster 2002). Due to the forward looking nature of the Foresight process the effect is thought to be more than just networking as such.

It is reckoned that through Foresight activities actors develop a better awareness of future risks and opportunities and a stronger inclination towards long term strategic thinking and better access to relevant knowledge for developing their strategic planning. This way, Foresight contributes to an infrastructure of "distributed intelligence" that is enabling the whole system to better address future challenges (Kuhlmann 2001).

2. Foresight orienting innovation towards societal needs

Besides improving the general system capability there is a more particular claim that Foresight can increase the quality of innovation processes by linking science and technology more closely with societal demands (Martin and Johnston 1998). By offering a forum for exchange between demand and supply perspective, it is argued, Foresight can orient innovation towards societal needs and future users' demands in an early phase of the innovation trajectory (Salo and Cuhls 2003).

3. Foresight providing information as a base for decision making

The idea of Foresight as a systemic innovation policy instrument is drawing on benefits that mainly arise from the Foresight process whereas the actual product that is generated within this process such as a report or even a recommendation is more of a means to structure this process. In line with this, the so called "process benefits" have more and more been emphasised by Foresight practitioners.

However, this does not imply that the actual provision of anticipatory intelligence to policy makers to inform their decision making is becoming less important. Foresight is still aiming to generate information for policy makers on possible pathways for the future and policy measures needed to foster these pathways. A solid set of methods is available in Foresight to generate this kind of intelligence based on the wide range of diverse knowledge of the Foresight participants. With respect to technological developments it is often elaborated in Foresight reports how certain technological developments may impact on the society and economy (exploratory Foresight) or the other way round how certain desired objectives can be reached through certain technology developments (normative approach). It is reckoned that due to the high diversity of knowledge sources that can be mobilised in a participatory Foresight process the quality of this information is better than just narrow expert advice. Often this type of information is directed at guiding research policy in priority setting for R&D funding. Typical products are scenarios and roadmaps describing future technology in society. In some exercises explicit policy recommendations are produced on how to reach the desired scenarios.

4. Foresight as agenda setting

In a number of Foresight exercises visions are elaborated describing future states of the domain that is tackled by the exercise. On the one hand these visions are just used as a means to derive anticipatory intelligence as outlined above. However they are also meant to have a function on their own: they evoke certain expectations towards a technology and thereby motivate actors to mobilise resources and invest in certain technological trajectories. For policy makers and other innovation actors such as engineers and managers they can become guiding visions that influence the way a technology is perceived designed and framed thus influencing the further evolvement of the trajectory. It is reckoned that within the networks formed by Foresight a common understanding of a certain future challenge emerges and some actors even orient their actions towards joint visions (soft coordination).

6.2.3 Summary

We have been discussing 4 main ways Foresight may impact on technological innovation processes or help policy-makers to do so:

1. Foster innovation capability (systemic instrument)
2. Oriented innovation towards societal & user demands
3. Inform policy makers on possible socio-economic implication of technological trajectories
4. Agenda setting across relevant actor groups

Finally it should be emphasised that most Foresight exercises follow a mix of these strategies to impact on technology policy. However, the emphasis and depending on it the approach is widely varying.

6.3 Social Shaping of Technology

There is a rich body of research results from various disciplines investigating the complex relationship between technology and society that has become known as SST (social shaping of technology). SST is characterised by an insistence that the 'black-box' of technology must be opened, to allow the socio-economic patterns embedded in both the content of technologies and the processes of innovation to be exposed and analysed (MacKenzie and Wajcman 1985; Bijker and Law 1992). In this sense it emerged through a critique of 'technological determinism'.

The insights from SST have gained increasing recognition in recent years as a valuable research focus, for its broader significance for the scientific and policy claims of social sciences. SST has offered a greater understanding of the relationship between scientific excellence, technological innovation and economic and social well-being; and in broadening the policy agenda, for example in the management of technological change and innovation.

SST studies show that technology does not develop according to an inner technical logic but is instead a social product, patterned by the conditions of its creation and use. Every stage in the generation and implementation of new technologies involves a set of choices between different technical options. Alongside narrowly 'technical' considerations, a range of 'social' factors affect which options are selected – thus influencing the content of technologies, and their social implications.

6.3.1 Co-Evolution of Society and Technology

As mentioned, the point of departure for SST research is the rejection of "technological determinism" that is the conceptualisation of technology as a phenomenon that is external to society and developing out of an own inner logic. While the opposition to technological determinism is a common denominator for SST studies the range of perspectives and approaches as well as disciplines is wide. However, across the diverse approaches a common understanding has emerged of technological change as a continuous process of socio-technical reconfiguration without any predetermined dominance of either material artefacts or social structures. Technological change and social change are analysed as being deeply intertwined and not to be isolated from each other. Many studies have highlighted the complexity of this co-evolution leading to the emergence of certain socio-technical patterns. SST research groups have focused on different aspects in their analysis of co-evolution.

Studies from a background in evolutionary economics rather start from the observation of certain patterns of technology development on a macro level. These scholars interpret the emergence of technological trajectories as a process of variation and selection with the selection environment being formed by socio-technical regimes on various levels (Geels 2004). As a result of this process it is argued, some new ideas develop into more stable technological trajectories. With increasing integration into socio-economic framework and embedding into institutional change these trajectories gain momentum and cannot easily redirected (path dependency). In order to better understand the dynamics of this process, investigations from this group of researchers have been aiming to closer define the interplay between variation and selection on various levels. Within STS, evolutionary economists have influenced the discipline during the 1980s using concepts like trajectories and regimes (Dosi 1982). The utility of using a particular technology increases with the number of adopters (the network effects). Therefore, stabilisation can emerge spontaneously, leading to the "lock-in" of one technology. This phenomenon has been related to the emergence of dominant designs in the history of various industries.

One well-known example of a "lock-in" is the QWERTY keyboard (David 1985). This keyboard was engineered in order to optimize typing speed in the case of mechanical typewriting. It had been designed so that the type bars have a minimum chance of jamming given the character frequency distribution in the English language. Since mechanical typewriting is out of use, the QWERTY keyboard has become suboptimal. However, one is no longer able to break out of the lock-in given learning curves and network externalities (Arthur 1988).

Another group of researches from the SST field has investigated the activities of innovation actors actively trying to shape the co-evolution process by weaving heterogeneous networks of material and social elements (Law and Callon 1992). Others have described how in early phases of innovation processes relevant social groups with their different interests negotiate the interpretation of new artefacts thereby continuously opening up new socio-technical options until finally a kind of closure is achieved (Bijker et al. 1987). The importance of power of social groups in defining this process has been highlighted (Hård 1993; Woolgar 1991).

The SST field is characterised by a number of detailed empirical case studies of past and current technological pathways elaborated by historians or sociologists of technology and scholars from a number of other disciplines. Based on these cases a number of concepts have been proposed to understand the emergence of certain technological trajectories. To collect insights relevant for the impact of Foresight on technological innovation we will pick up upon on some basic insights from SST research in more detail: Contingency and constraint of variation, Role of expectations and visions, Importance of downstream phase of innovation, Importance of localisation and Insights on steering possibilities for technological trajectories.

6.3.2 Flexibility and Contingency Vs. Constraints in the Design Phase

The co-evolution process is at the same time contingent in the sense that it could evolve different depending on the context but on the other hand structured through the embedding in an existing framework. This seems to be of special interest for the early phases of technological trajectories that is often targeted by R&D policy and Foresight activities. Some SST studies have focused on the phase of invention and emphasised the interpretative flexibility and contingency of the technology design process. On the other hand the way design is structured by its embedding into existing regimes (e.g. firm routines, best practice rules etc.) has been highlighted.

For example Williams and Edge (1996) describe how, in the IT sector we find many complex patterns of stabilisation and destabilisation. The emergence of industry standard products (black-boxed solutions) 'creates' markets. These offer cheaper products, and give users both a greater choice of suppliers, and confidence that a product will not become obsolete (Swann 1990). This creates an incentive for suppliers to collaborate in creating larger and more stable markets. Increasingly, firms are coming together, with competitors and suppliers of complementary products, to agree standards for emerging technologies (Cowan 1992; Collinson 1993). Future technologies/markets are being pre-constructed in a virtual space constituted by the collective activities of players. However there is not, of course, a unidirectional shift away from competition. For example, these markets may attract new entrants (e.g. the proliferation of vendors of IBM pc 'clones'). Where accommodation or collaboration is not favourable, firms may promote proprietary solutions. Dominant players may seek to destabilise solutions and erode industry standards, to monopolise their links with users – for example the recent, largely unsuccessful, attempt by IBM to tie in existing users to their next generation of personal computers by launching the new OS2 operating system in place of the industry standard DOS.

6.3.3 Role of Visions and Expectations

Expectations or visions about technologies can be defined as "real time representations of future technological situations and capabilities" (Borup et al. 2006). From early on, SST studies have investigated the role of visions and expectations in shaping technological trajectories. Different perspectives have been taken.

On a more micro level it has been observed how collective visions of users and use of technical artefacts held by engineers and technicians influence design decisions and, once embodied in the artefact, structure later users' possibilities of use (Konrad 2004; Akrich 1992; Woolgar 1991). It has been stressed how such visions are closely tied to specific social experience of actor groups.

A growing number of studies in SST are tackling the role of collective expectations in shaping technological innovation on a more meso and macro level. Often, emerging technologies are associated with certain benefits such as positive impact on quality of life, environment or economy. Expectations that are shared among a group of actors in a certain domain have been shown to be of considerable impact on the technological trajectory. Studies with an evolutionary economics background have placed the dynamics of expectations as a central interface between various levels of the selection environment determining technological trajectories. So e.g. due to positive expectations niches are granted where new technologies can be developed in a protected space and learning between developers and users takes place. Within the niche the new socio-technical configuration can stabilise and later modify the wider regimes on meso or even macro level leading to a regime change or even transition (Kemp 1994). Thereby on all levels of the innovation process visions and expectations play a performative role. The reason for their strong influence is the ability to bridge between different groups of actors such as policy-makers and research community, managers and scientists, users and providers of innovation (Borup et al. 2006). This does not imply that expectations are always fulfilled. On the contrary many of them fail because of too naive and linear projections (Geels and Smit 2000). Nevertheless the existence of the expectations directs the innovation activities as promises and expectations are translated requirements for further technology development. The whole process has been described as a "promise-requirement cycle" (Van Lente 1993).

6.3.4 Importance of Downstream Phase of Innovation

A great deal of SST research has looked into later stages of the innovation process such as diffusion, adoption, consumption of technological artefacts and their embedding into organisational settings. Many SST studies highlight how through appropriation by a socio-cultural context technologies are reshaped and redefined. Cultural studies have particularly emphasised the crucial role of domestication of technology, i.e. incorporation into daily life routines and assignment of symbolic meanings. The role of users and ways of using in shaping this change has been one of the core lines of investigation of SST scholars (Woolgar 1991; Kline and Pinch 1996; Oudshoorn and Pinch 2003). Results from the many SST case studies investigating current and past cases of technological innovation clearly indicate that users indeed play a major role in the shaping of technology: For a number of technological innovations it has been shown how unexpected forms of use can redefine the concept of technological artefacts and how competing expectations and interpretations from different user groups influence the direction of technological pathways (Hård 2002). Even a successful product is re-contextualised in different ways several times during its innovation trajectory. Accordingly, it has been concluded that the more downstream phases of innovation should be fully considered as part of the innovation cycle. Thus SST research has strongly underlined the move away from linear models of innovation featuring successive phases from invention to consumption. Other terms such as "Innovation journey" or "distributed innovation" have been proposed to emphasise the meandering path of an innovation process (Rip and Schot 2002).

In their book How users Matter, Oudshoorn and Pinch (2003) provide some provocative examples of the co-construction of users and technology; "New uses are always being found for familiar technologies. Sometimes these changes in use are dramatic and unexpected. Before September 11, 2001, no one foresaw that an airliner could be turned by a small number of its occupants into a giant Molotov cocktail".

6.3.5 Importance of Localisation

SST studies have found that the interplay between universal and local elements is crucial to understand socio-technical change (see e.g. Hård 1994). Although certain patterns of technological development can be outlined on a macro level, the real process is always characterised by interplay between local configuration activities and universal patterns. Thus the same technology might be incorporated into different socio-technical patterns in different socio-cultural contexts. And on the other hand socio-technical patterns that have been developed on a local level need to be translated before they can feed into universal patterns forming the structuring regime for again other innovation activities.

Especially studies from the so called "actor-network approach" remain sceptical about the nature and influence of broader social and economic structures of power and interests, insisting that actors create the world anew (Latour 1983, 1986, 1988), and implying that technologies (and social systems generally) are highly malleable to local actors.

6.3.6 Strategies for Intervention

A number of SST scholars have explicitly drawn conclusions from their results as to the possibility of intervention into socio-technical transformation processes (Sørensen and Williams 2002). As a first result it needs to be emphasised that the concept of the "social shaping" of technology does not imply easy steering of technological trajectories. On the contrary, the deep embedding of technological elements into socio-technical arrangements implies a high degree of stability and a certain resistance to change. Socio-technological trajectories once emerged carry a momentum causing a certain path-dependency so they cannot be easily redirected. In particular direct head-on intervention is likely to fail or cause unintended consequences. However, the detailed insights into the complex process of socio-technical change allow some different kind of intelligent intervention that may effect a modulation of trajectories (Rip and Schot 2002).

The following modes of intervention for policy have been suggested by SST scholars:

- Support to formulation of socio-technical scenarios in an early stage of emerging technologies (Rip and Schot 2002) to add other interests (e.g. policy goals) to promote diversity of visions into the promise-requirement cycle, (Russel and Williams 2002)
- Provision of protected spaces (niches) for social learning about new technologies first on a micro level and later on a meso level e.g. through social experiments
- Combating entrenchment and early closing by providing wider arenas and constant monitoring (Russel and Williams 2002, p. 54)
- Providing communication channels (Russel and Williams 2002, p. 54) and support the forming of new alliances and networks thereby facilitate alignment of elements into new socio-technical configurations
- Foster dialogue between users and providers of innovation
- Target structuring regimes in the generation of knowledge (e.g. engineering education)

6.4 Some Implications for Foresight

What are the lessons to be learned from the SST results on the possible impact of Foresight on the innovation process? As discussed above, Foresight is aiming at such an impact on the one hand through providing intelligence as a base for decision making and on the other through moderating processes that enhance the responsiveness of the arena. We consider that for both types of impact Foresight insights from SST can be exploited. The following main inroads for SST on Foresight concept and practice can be highlighted:

6.4.1 Foresight as Process Moderator

The relevance of a systemic innovation policy instrument that is working on the process aspect of innovation by establishing networks and providing spaces for mutual learning is clearly confirmed by SST results. Indeed SST research implies that this type of intervention might be even more effective than classical measures targeting demand or supply side in isolation or attempt to intervene more directly. In particular SST results confirm that the quality of any innovation system is likely to benefit greatly from the provision of joint learning spaces between users and producers of innovation.

However, the SST insights do also suggest some issues to beware of. The complexity and contingency of socio-technical change as emphasised by SST implies that the selection of actors to be involved in the process is not at all a straightforward one. A careful mapping of the arena of change taking into account a wide range of possible trajectories of change and especially the downstream parts is needed. It cannot be assumed that central actors themselves stemming from either supply or demand side organisations such as companies or associations that are typically involved into a Foresight process have an adequate overview of this space. The involvement of social scientists into the stakeholder mapping phase of Foresight might be very useful. Foresight that is just working with the "usual suspects" is in danger of even increasing lock-in situations into less desirable trajectories. The creation of diversity is important needs to be systematically targeted through the design of the exercise (Könnölä et al forthcoming).

6.4.2 *Foresight as Expectation Management*

SST results clearly confirm the relevance of expectations and visions in directing technological trajectories especially in an early phase. SST researchers have highlighted the support to vision building as one of the possible loci of policy intervention. However SST research also indicates that not any kind of vision is suitable to modulate innovation trajectories towards policy goals. On the contrary, narrow visions that are pushed out of interest of only a limited range of actors might become a barrier to flexibility and openness of innovation towards various societal demands. The primary concern of Foresight should therefore be a diversification of expectations and visions.

Also, visions that project technological developments into the future in a linear way without taking into account the complexity of the innovation journey and especially the appropriation phase of innovation cycle are not likely to be of use to policy strategy building. Possible ways of policy intervention on the demand side do not become visible if the uptake of a certain technology is just taken for granted in a vision. At the same time, possible barriers grounded in the process of appropriation might well be overlooked. This means that visions and expectations do indeed matter and provide an important inroad to influencing technological trajectories but to be useful they need to be as rich on the side of society as on technology. To actually develop such meaningful socio-technical future visions Foresight needs to look at societal development and technological possibilities with the same degree of openness and expertise. This again demands expertise on the dynamics of social change just as much as expert knowledge about technologies. Again, social scientists will have a contribution to make here. Furthermore, the involvement of the relevant social groups (e.g. potential user groups) themselves will greatly improve the quality and usefulness of the visions developed.

Finally, a suggestion might be derived from SST results with respect to the format of visions developed in Foresight. As SST has shown, expectations and visions will never be fully met. Instead within the "promise requirement cycle" (Van Lente 1993) visions are continuously adapted along the line. Nevertheless their function is to motivate resource allocation but also to orient experimenting. This means that they need not necessarily be exhaustive in the sense that they describe a whole set of possible socio-economic frameworks around the "new technology". Instead, it might sometimes be more relevant for the visions to be rich in suggestive detail and contain a number of imaginative thought provoking elements. A set of small narratives or scripts describing possible future socio-technical ensembles might sometimes be useful to replace or at least complement large scale scenarios if the aim is to modulate technological trajectories).

6.4.3 Provision of Anticipatory Intelligence

Whereas for visions diversity and creativity are crucial "anticipatory intelligence" that is used as a base for policy decision making is bound to give insights that can be operationalised into policy action. Therefore it is important that it takes into account the real world processes of technological change. Anticipatory intelligence therefore needs to focus on co-evolution processes as described by STS research. Socio-technical scenarios giving realistic descriptions of use and embedding into socio-cultural context or various organisational settings can give valuable insights on possible points of intervention. However, also other Foresight methods are able to generate knowledge about co-evolution provided that this is carefully targeted (van der Meulen 2003) e.g. by considering appropriation processes as diligently as technological developments. In such a way reflexive anticipation is likely to allow for the identification of possible pathways for transition towards desired trajectories as well as the design of adequate process oriented policy measures (Geels 2002a,b).

6.4.4 Localisation Through Foresight

SST results have shown how local contextualisation is an important element in the forming of technological trajectories. This implies that different innovation systems will each incorporate technological elements in a different way. For Foresight this implies that local level analysis e.g. for a region or city should not just take over visions and pathways from higher levels. The embedding of an innovation into a specific regional setting should itself be considered as part of the innovation process and not just a further "adoption" with specific regional consequences. The same holds for various industrial sectors that will each have to go through their own innovation cycle when taking up new technologies.

6.5 Conclusions

In this paper it was explored how results from social science research on the social shaping of technology can be exploited for Foresight concept and practice. As a first result it was emphasised that the approach of Foresight to foster innovation capability by initiating a collective learning and vision building process is well in line with SST results. In fact SST research suggests that systemic or process oriented instruments such as Foresight are one of the most likely to impact on the complex interplay of factors governing innovation trajectories. It was furthermore suggested that for innovation policy the creative and visionary aspect of Foresight might even be more relevant than often recognised by Foresight practitioners and users, as SST results highlight the decisive impact of visions and expectations on the innovation process.

We have illustrated how technologies are complex constructs that can be appreciated differently from various perspectives. In general, three main subdynamics have been distinguished: the selection mechanism associated with the market and society at large, the historical generation of technological variation along the time axis, and the wish to control the technological development from a government perspective. These perspectives have different meanings in the context of relevant theoretical traditions (Leydesdorff 1997). Each of the perspectives reduces the complexity by taking a specific angle. When a single perspective is dominant (for example, that of a central state or short-term profit taking), the technological development can be blocked temporarily. Both governments and innovators play a role in changing the technological developments. The lack of closure of technological developments challenges our inventiveness and thereby it liberates the forces of technological innovation for the creative destruction of previous production relations. The desired role of the government is to provide the interface of networking between the actors, to develop a better awareness of future risks and opportunities and a stronger inclination towards long term strategic thinking and better access to relevant knowledge for developing their strategic planning. This way, Foresight contributes to an infrastructure of "distributed intelligence" that is enabling the whole system to better address future challenges (Kuhlmann 2001).

However, cautioning insights from SST towards Foresight practice were also highlighted. It was pointed out that in order to impact on the complex co-evolution process of society and technology, Foresight needs to tackle the social dynamics of change more diligently. In particular Foresight visions need to better reflect the complexity of societal change by taking into account e.g. unexpected forms of social approbation of technology. It was suggested that in some cases there might be a need for more "realistic" visions or scripts of future forms of use of technological artefacts to influence expectations and less for large scale scenarios of socio-technical transformation. Finally it was stressed that when initiating a Technology Foresight process the arena of change needs to be mapped very carefully for relevant actors and stakeholders. Especially on the more downstream side of the innovation process, relevant actor groups need to be considered. The involvement of social scientists into Foresight design seems likely to be useful to fulfill both these demands. To further exploit SST for Foresight the next step will be to have a look at real Foresight exercises and see how the lessons learned from the theoretical considerations can be translated into Foresight practice.

Finally, we would like to point out some issues worthwhile to explore further in the context of this debate. It seems that the difficulty of Foresight to adopt a more holistic view on socio-technical change might partly be rooted within the structure of the policy arena. Foresight exercises are often financed by R&D departments to decide on R&D priorities with little possibility to act on the wider socio-economic framework. In these cases the challenge for Foresight will be to interpret insights stemming from the holistic analysis of socio-technical co-evolution back into conclusions for modes and content of R&D funding.

While here the attempt has been to make Foresight results more useful for innovation policy by integrating better understanding of real innovation processes it is important to keep in mind that it is still another thing to actually orient innovation towards normative policy objectives such as sustainability or quality of life. There is no guarantee that wiring up innovation systems and even introducing a better user orientation will automatically lead towards such desired directions (Weber 2006). Other approaches such as strategic niche management (Kemp et al. 1998; Weber 2006) or transition management (Rotmans et al. 2001) will have to be employed if such an impact is desired.

References

M. Akrich, (1992) Beyond social construction of technology: The shaping of people and things in the innovation process, in M. Dierkes and U. Hoffmann (eds.) New Technology at the Outset: Social Forces in the Shaping of Technological Innovations, Campus/Westview, Frankfurt/NY, Chapter 9, pp. 173–190

W.B. Arthur,(1988) Competing technologies: An overview, in G.C. Dosi, R. Freeman, N.G. Silverberg and L. Soete (eds.) Technical Change and Economic Theory, Pinter, London

W.E. Bijker, T.P. Hughes, T.J. Pinch (eds.) (1987) The social construction of technological systems – New Directions in the Sociology and History of Technology, Cambridge, MA

W. Bijker, J. Law (eds.) (1992) Shaping Technology/Building Society: Studies in Socio-Technical Change, MIT, Cambridge, MA, London

M. Borup, N. Brown, K. Konrad, H. Van Lente (2006) The Sociology of Expectations in Science and Technology. Technology Analysis & Stategic Management, Vol. 18, No. 3/4, pp. 285–298, July–September

S. Collinson (1993) Managing product innovation at Sony: the development of the Data Discman. Technology Analysis and Strategic Management, Vol. 5, No. 3, pp. 285–306

R. Cowan (1992) High technology and the economics of standardization Chap. 14, pp. 279–300, in M. Dierkes and U. Hoffmann, (eds.), New Technologies at the Outset – Social Forces in the Shaping of Technological Innovations, Campus/Westview

A.P. David (1985) Clio and the Economics of QWERTY. American Economic Review, Vol. 75, pp. 332–337
G. Dosi (1982) Technological Paradigms and Technological Trajectories: A Suggested Interpretation of the Determinants of Technological Change. Research Policy, Vol. 11, pp. 147–162
F.W. Geels (2004) Processes and Patterns in Transitions and System Innovations: Refining the Co-Evolutionary Multi-Level Perspective. Technological Forecasting & Social Change, Vol. 72, pp. 681–696
F.W. Geels (2002a) Technological Transitions as Evolutionary Reconfiguration Process: A Multi Level Perspective and a Case Study. Research Policy, Vol. 31, pp. 1257–1274
F.W. Geels (2002b) Towards sociotechnical scenarios and reflexive anticipation: using patterns and regularities in technology dynamics. in KH Sørensen and R Williams (eds.) Shaping Technology, Guiding Policy: Concepts, Spaces & Tools, Edward Elgar, Cheltenham, UK, pp. 359–385
F.W. Geels, W.A. Smit (2000) Failed Technology Futures: Pitfalls and Lessons from a Historical Survey. Future, Vol. 32, pp. 867–885
L. Georghiou (2001) Third generation foresight – integrating the socio-economic dimension, In Technology Foresight – The Approach to and Potential for New Technology Foresight, Proceedings of the International Conference on technology Foresight, NISTEP Research Material, vol. 77 2001 (March)
M. Hård (1993) Beyond Harmony and Consensus: A Social Conflict Approach to Technology Science. Technology & Human Values, Vol. 18, No. 4 (Autumn), pp. 408–432
M. Hård (1994) Technology as Practice: Local and Global Closure Processes in Diesel Engine Design. Social Studies of Science, Vol. 24, No. 3 (August), pp. 549–585
M. Hård (2002) Cultural Politics in Action: Developing User Scripts in Relation to the Electric Vehicle. Science Technology and Human Values, Vol. 27 No. 2, pp. 262–281
A. Havas (2005) Terminology and Methodology for Benchmarking Foresight Programmes, Prepared for the For Society project, Tasks 1.4 and 3.1
S. Jasanoff, G.E. Markle, J.C. Petersen and T. Pinch, (eds.) (1994) Handbook of Science and Technology Studies, Sage Publications, Thousand Oaks, London & New Delhi
R, Kemp (1994) Technology and the Transition to Environmental Sustainability: the Problem of Technological Regime Shifts. Futures, Vol. 26, pp. 1023–1046
R. Kemp, J. Schot, R. Hoogma, (1998) Regime Shifts to Sustainability through Processes of Niche Formation: The Approach of Strategic Niche Management. Technology Analysis and Strategic Management, Vol. 10, 175–195
K. Konrad, (2004) Prägende Erwartungen. Szenarien als Schrittmacher der Technikentwicklung edition sigma, Berlin
T. Könnölä, V. Brummer, A. Salo, (2006) Diversity in Foresight: Insights from the fostering of Innovation Ideas, Helsinki University of Technology, Systems Analysis Laboratory, Research Report E19
R. Kline, T. Pinch (1996) Users as Agents of Technological Change: The Social Construction of the Automobile in the Rural United States. Technology & Culture, Vol. 37 No. 4, pp. 763–795
S. Kuhlmann (2001) Management of Innovation Systems the role of Distributed intelligence, Maklu-Uitgevers, Antwer
B. Latour (1983) Give me a laboratory and i will raise the World, in Knorr-Cetina, M. & Mulkay, M. (eds.), Science Observed, Sage, London, pp. 141–70
B. Latour (1986) Science in Action, Open University Press, Milton Keynes
B. Latour (1988) How to write "The Prince" for Machines as well as Machinations, in B. Elliot (ed.) Technology and Social Change, Edinburgh University Press, Edinburgh, pp. 20–43
J. Law, M. Callon, (1992) The life and death of an aircraft: a network analysis of technological change, in Bijker, W. and Law, J. (eds.) Shaping Technology/Building Society: Studies in Socio-Technical Change, MIT, Cambridge, MA, London, pp. 29–52
L. Leydesdorff (1997) The Non-linear Dynamics of Sociological Reflections. International Sociology, Vol. 12, pp. 25–45

B-Å Lundvall, S Borrás (1998): The Globalising Learning Economy: Implications for Innovation Policy Brussels: Commission of the EU. (170 pages). http://www.cordis.lu/tser/src/globec.htm

D. MacKenzie, J. Wajcman, (eds.) (1985) The Social Shaping of Technology: How the Refrigerator Got Its HUM, Open University Press, Milton Keynes

B.R. Martin, R. Johnston (1999) Technology Foresight for Wiring Up the National Innovation System. Technological Forecasting and Social Change, Vol. 60, pp. 37–54

N. Oudshoorn, T. Pinch, (2003) (eds.) How Users matter: The Co-Construction of Users and Technology, MIT, Cambridge, MA, London

A. Rip, J.W. Schot (2002) Identifying loci for influencing the dynamics of technological development, in K.H. Sorensen, R. Williams (eds.) Shaping Technology, Guiding Policy: Concepts, Spaces & Tools, Edward Elgar, Cheltenham, UK pp.155–172

J. Rotmans, R. Kemp, M. van Asselt (2001) More Evolution than Revolution. Transition Management in Public Policy. Foresight, Vol. 3, No. 1, pp. 15–31

S. Russel, R. Williams (2002) Social shaping of technology: frameworks, findings and implications for policy, in K.H. Sørensen, R. Williams (eds.) Shaping Technology, Guiding Policy: Concepts, Spaces & Tools, Edward Elgar, Cheltenham, UK

A. Salo, K. Cuhls (2003) Technology Foresight – Past and Future. Journal of Forecasting, Vol. 22, pp. 79–82

R. Smits, S. Kuhlmann (2004) The Rise of Systemic Instruments in Innovation Policy. International Journal of Foresight and Innovation Policy, Vol. 1 No. 1/2, pp. 4–32

K.H. Sørensen, R. Williams (eds.) (2002) Shaping Technology, Guiding Policy: Concepts, Spaces & Tools, Edward Elgar, Cheltenham, UK

P. Swann (1990) Standards and the growth of a software network, in J.L. Berg, H. Schumny (eds.) An Analysis of the Information Technology Standardization Process, Elsevier Science/North-Holland, Amsterdam, pp. 383–393

A. Tubke, K, Ducatel, J. Gavigan, P. Moncada (eds.) (2001) Strategic Policy Intelligence: Current Trends, the State of Play and Perspectives, IPTS, Seville

B. van der Meulen (2003) Integrating technological and societal aspects of ICT in Foresight Exercises Technikfolgenabschätzung – Theorie und Praxis Nr. 2, 12. Jg., Juni 2003 pp. 66–74

H. Van Lente (1993) Promising Technology – The dynamics of Expectations in technological Developments, Eburon, Enschede

M Weber (2006) Foresight and Adaptive Planning as Complementary Elements in Anticipatory Policy-making: A Conceptual and Methodological Approach, in J.-P. Voß, D. Bauknecht, R. Kemp (eds.) Reflexive Governance for Sustainable Development, Edward Elgar, Cheltenham, pp. 189–222

A. Webster (2002) Response to Prof. Remi Barre's Foresight and Their Themes, Paper presented at The Role of Foresight in the Selection of Research Priorities hosted by the Spanish Ministry of Science and Technology, May

R. Williams, D. Edge (1996) The Social Shaping of Technology. Research Policy, Vol. 25, pp. 856–899

S. Woolgar (1991) Configuring the user: The case of usability trials. In J. Law (ed.) A Sociology of Monsters: Essays on Power, Technology and Domination, Routledge, London

Chapter 7
Strategic Intelligence in Decision Making

P. De Smedt

7.1 Introduction

There are many reasons why strategic intelligence is required to support policy decisions. These primarily stem from the nature of today's knowledge society with two contrasting trends. On one hand, there is a trend of increasing human intelligence in the economic, social and political systems (Hughes 2007). On the other hand, there is a trend towards dissolving certainties about the problems and solutions of today's society (Hoijer et al. 2006). Clearly, more information does not always imply more certainties on how to act and even the same facts are often interpreted in markedly different ways. The same policy relevant information can and often does result in conflicting framing of a problem by different stakeholders. This is rather because of competing assumptions than because of inconsistent facts (Dunn 2004). Therefore, it is not surprising us that policy-makers are demanding for strategic intelligence to support their understanding of today's challenges, including the relevant aspects and impact of science and technology and their possible future developments.

Strategic Intelligence (SI) applications – such as forecast, impact assessment and foresight exercises – have been developed to support decision-making. Examples demonstrating the diversity and broad application of SI, can be found in a wide range of scientific literature, in project reports and occasionally in policy documents. Still, limited information can be found on the impact, limitations and effectiveness of SI applications.

This chapter on strategic intelligence in decision-making reflects on policy analysis concepts, such as the evidence-based approach and the rational decision-making model, and explores the core problems concerning the effectiveness of SI applications to support decision-making. The hypothesis is that SI applications need to be better institutional embedded in terms of opportunity, purpose and legitimacy, so that SI applications do not become meaningless and useless for the decision-makers.

The first part (Sects. 7.2–7.4) looks at different concepts of decision-making, including evidence for policies, the role of politicians and policy change. The first section lists witch types of evidence for policy can be distinguished, who can provide

C. Cagnin et al. (eds.) *Future-Oriented Technology Analysis*,
doi:10.1007/978-3-540-68811-2, © Springer-Verlag Berlin Heidelberg 2008

the knowledge, and how the best evidence can be mobilised. The second section describes the different positions a politician can take in the policy process and also looks in at decision failure. The third section defines policy as a process and explores the nature of policies and the dynamics of policy change.

In the second part (Sects. 7.5 and 7.6) three complementary perspectives are proposed to analyse the effectiveness of SI applications in decision-making. The three complementary perspectives – window of opportunity, clarity of purpose, and legitimacy of evidence – were also applied to analyse a foresight exercise with scenario workshops.

In the third and final part, main conclusions on the interaction between decision-making and SI applications are described, offering insights to improve policy practice. The proposed analytical framework can also be used as a research agenda developed to improve the theoretical underpinnings of SI applications.

7.2 Evidence for Policy

A policy is a deliberate plan of action, guiding decisions and achieving rational outcomes. From a strategic perspective, the role of a policy is to resolve contradictions between the organisation and its environment. Broadly, policies are typically instituted in order to seek positive benefit and to avoid negative effects. The goals can vary widely according to the context. The purpose is not simply to provide a basis for making efficient decisions, but also to provide knowledge needed to improve the organisational, political and social systems. The overall idea is to promote human development by reasoning how to achieve an improved society (Dunn 2004). Though, policies can also have side effects or unintended consequences. For example, a government may make a policy decision to raise taxes in the hope of increasing overall tax revenue. But, depending on the context and the size of tax increase, this may have the effect of reducing tax revenue by causing capital flight (Dunn 2004).

7.2.1 Types of Policy Knowledge

The notion of evidence-based policy fits well with a rational decision-making model (Davies et al. 2000). The solution of a complex social problem requires not only better evidence of what works in terms of policy intervention, but also requires more rational decision-making in which such evidence can play a stronger role (Sanderson 2004). Colebatch (2006) describes tree types of policy knowledge (based on Tenbensel 2006):

Epistemic knowledge: the universal knowledge produced by analytic rationality. Epistemic knowledge is the type that establishes causal links and chains, and is the knowledge aspired to by mainstream rationalist policy analysts in their search for the likely consequences of the different policy alternatives they evaluate.

Tacit knowledge: the practical-technical knowledge derived from experience and skill. This is not simply the practical applications of epistemic knowledge. The tacit knowledge rests very much in implicit personal or institutional practices often associated with craft like skills, awareness of reputations, hands on techniques, etc. It is the knowledge which cannot be explicitly codified.

Phronetic knowledge: this is a sense of the ethical. It is based on practical value rationality. 'Where are we going?', 'Is this desirable?', and 'What should be done?' are phronetic questions. This type of knowledge is important because it is often needed to underpin the definition of a policy problem.

The point here is that policy arguments are likely to involve all of these sorts of knowledge, but that participants are unlikely to be equally skilled in all of them. The reason for this is because each type of knowledge asks a different question. Episteme asks 'what is true?'; tacit knowledge asks 'what works?; and phronetic asks 'what should be done?' Good policy argument rests on a foundation of all three types of knowledge (Colebatch 2006).

7.2.2 Providers of Policy Knowledge

In many policy areas participation of actors from society has become common practice. Participation can take place in different forms and at different levels. For instance, stakeholders can be involved in local planning processes in which they have actual decision-making authority. They can also operate as advisers by sitting on national or regional boards of decision-making bodies. In all cases, the involvement of stakeholders is expected to deliver a useful contribution to the policy-making process because of different reasons. Stakeholder participation can for instance help to mobilize specific expertise that these actors have, it can improve awareness and support for specific policy measures; it can enhance the legitimacy of the decisions taken, and can also build new policy networks and coalitions to support cooperation on the long term (van de Kerkhof 2006).

A general definition of public participation is the practice of involving members of the public in the agenda setting, decision-making, and policy forming activities of organizations responsible for policy development (Rowe and Freyer 2005). The stakeholders are members of the public who own the problem under discussion and having a stake in the future. Stakeholders can be individuals, informal groups or well established organisations. The number of stakeholders involved in a certain issue is not necessarily fixed but might change over time. As the policy process evolves, new stakeholders will enter the scene and others will leave.

Stakeholder participation is considered to be a key driver behind improving evidence for policy (Enserink 2003). Decisions will be better in two respects: first, they will command greater respect from the stakeholders involved and hence carry more legitimacy; and second, they will benefit from the insights and knowledge brought by the different stakeholders (Burton et al. 2006). This used to be the domain of scientists and technical experts and reflects the changing views on science-policy interface (van den Hove 2007). Policy research also reveals that, whilst being a form of inclusiveness, participation can have significant deficiencies. This is because in much practice, such as Delphi exercises, it presents the majority view and it is therefore limited with respect to the range of values and preferences that it can elaborate. Practice also points out that greater stakeholder participation makes the policy process more complicated, complex and unpredictable. Whilst we might be confident that these policy processes will be different, we cannot precisely know in advance what these differences due to greater participation will be (Burton et al. 2006).

7.2.3 Mobilising the Best Available Evidence

Society is a complex system of interacting elements and is influenced by the external environment. In order to assess and resolve organisation issues, politicians and policy analysts create abstractions of these systems in the form of mental models. Mental models can be described as the lenses through which people see the world. These models incorporate the biases, values, and beliefs of people. Mental models are highly subjective and are depending on the world-views, as well as on the historical knowledge regarding the situation. This implies that models are not descriptive for the real world, but rather descriptive for ways how to think about the real world. In a more managerial context, mental models are the ideas and conceptions about the organisation, how it has been and will be developing on the short term, and likely some ideas about where the organisation is heading for the long term.

Policy analysis or policy practice is a problem-solving discipline. It is distinct from pure academic research that is seeking theoretical knowledge. It is also distinct from a policy-orientated inquiry, characterised with a limited scope and mostly done to inform a specific decision. Policy practice supports decision making by identifying ways of thinking about society and policy change. It can be used to structure policy problems and to provide evidence underpinning decision-making. This recognition relates not only to Lasswell's belief in the importance of acquiring maximum rational judgment, but also to Hoppe's view that – in producing viable policy recommendations – policy practice should successfully mobilise the best available evidence in the desire to tackle problems on the political agenda (Geva-May 2002). Many have speculated that the key to improve decision-making lies in being conscious and aware about the nature of the mental model, i.e. being a description of perception rather then a description of reality (Schwartz 1991).

7.3 Decision Making

Decision-making is a cognitive process leading to a choice or a selection among variations. Decisions are social constructs. This means we can never see a decision. We can only observe the consequences. Decision-making can, and mostly is, influenced or supported by rational arguments and powerful forces of inertia, expediency, ideology and finance.

7.3.1 The Decision Makers

There are inherent tensions between traditional, more pluralist forms of public participation and new deliberative democratic processes. These innovative processes are challenging existing roles of the decision-makers in society. But the appreciation of these processes depends largely on the ingoing position taken towards the role of politicians in general. Hendriks (2002) makes an abstraction of a politician's role by describing two opposite positions: on one side a centralised and top-down steering approach, and on the other side a facilitating and networking approach. For each of these two opposite approaches he also describes a hard and soft approach.

Centralised: proponents of powerful politics are in favour of the classic notion of representative democracy. Politicians are elected representing the public interest and take precedence. The hard variant stands for a strong centralised leadership with a strong concentration of the decision power, while the soft variant tolerates more interactive consultation.

Decentralised: proponents of the modest role of politics are in favour of a facilitating role for politicians in decision-making. The hard variant stands for politicians who only steer and intervene when and if necessary, but who otherwise remain on the sidelines. The soft variant is more managerial oriented and in favour of politicians who are limiting their selves to network management: politicians as a creator of preconditions and rules of game, as a process facilitator.

The conceptual difference between stakeholders and decision-makers is clear. The former has a stake and can have an influence to the decision-makers. The latter has the responsibility and power to make the decisions. In practice policy decisions are often shrouded in uncertainty. It is not always clear that a decision has been taken. It is sometimes unclear what the decision is and who has taken it (Burton et al. 2006). In reality the role of a politician is dynamic and deviates depending on internal and external developments such as the actual political agenda setting, temporally coalitions, discontinuities and so one. The boundaries between stakeholders and decision-makers are less fixed and literature often includes also others non-politicians who are demonstrating political support to the process. Also agencies or government departments who are playing a role in the development of programs or in the allocation of funding can be seen as part of the decision-making, although this is sometimes contested.

7.3.2 The Risk of Decision Failure

Decision failure is more common than people often tend or wish to believe. Some decision-makers always expect good results ignoring the possibility that outcomes of a good decision may change. If a decision-maker gets caught up in decision failure, most often they reveal as little as possible. Additional it is also difficult to separate good decisions with bad outcomes from bad decisions with good outcomes. Anyhow, research on decision errors in organisations reveals high levels of failure, even up to fifty percent, in day to day decision-making (Nutt 2004).

Decision failures occur in two overarching categories: i. simple explainable errors or mistakes and ii. unexplainable or unexpected decision errors. The first category refers to the possibility that the decision-maker was unable to make the decision. This category of inevitable errors denotes the statistical necessity that some random error will occur. Decision failures that occur in the second category are more important because there is seemingly no logical explanation for the decision failure. The unexpected happened and the mental model turned out not to be robust enough (Chermack 2004). There are four potential contributors, each independently or combined contributing to decision failures, namely i. bounded rationality, ii. neglecting internal change, iii. stickiness and friction of information and knowledge, iv. mental models including decision premises or policies (see also Table 7.1).

Bounded rationality is related with the fact that information is endless but the translation into knowledge and appropriation by decision-makers is a struggle point. People do not have enough time to read, understand and synthesize the information on complex developments and their impacts.

Neglecting internal change refers to the given that decision-makers most often focus on external (exogenous) variables to anticipate. An explanation can be found in the fact that external variables are easily recognisable and some internal (endogenous) variables are more hidden in the system. Hence, a change in 'hidden' internal variables can have unforeseen consequences that become magnified because of their association with feedback processes.

Stickiness refers to a characteristic of information and is associated with the cost of its transfer between or among people. Friction can be described as the nuances and double-checks that occur in the social interactions in work processes. Frictionless knowledge would initially be more efficient and less sticky, but would also allow for a drastic increase in decision errors. The loss of friction will allow many errors to continue that were previously prevented during the course of social interaction.

Mental models are a concept that attempts to explain the way people frame their experience and from which they draw their assumptions about situations and

Table 7.1 Four contributors to decision failure (after Chermack 2004)

Bounded rationality	People cannot effectively cope with all of the available information and alternatives
Neglecting internal change	People have a tendency to believe that all internal processes are well being recognised
Stickiness and friction of information and knowledge	There are cost and limitations in the transfer of information and knowledge between people
Mental models	People are often selective and include and exclude information based on their mental model

alternatives. Often, people make decisions based on only a selection of the information and knowledge available. This is due to the fact that people include and exclude information based on their mental model.

7.4 Policy as a Process

The essential of policy work is very simple and all about choice: some people are making choices, and others are engaged in helping them to make the best choices (Colebatch 2006). In addition, most people often believe that policy problems are objective conditions, and that their solutions can be simply identified by determining what the facts are in a given case. This rather naïve view is in contrast with the experience of policy practitioners and fails to recognize the inherently ambiguity and complexity of policies today. In addition policy practitioners also highlight that evidence is regarded as a necessary, but not a sufficient, condition for any decision-making process. This is because knowledge is but one influence on the policy process and is not always influential, supplanted by be powerful political forces of inertia, expediency, ideology and finance (Sanderson 2004). The ways that policies are developed, implemented and revised are always shaped by the wider social and political contexts (Shaxson 2005).

7.4.1 The Nature of Policies

The dominant paradigm on policy practise sees the policy process as an exercise in informed problem-solving: a problem is identified, data is collected, the problem is analysed and advice is given to the policy-maker, who makes a decision when is then implemented (Colebatch 2006). This paradigm often seems to be at variance with the experience of policy practitioners. The same policy relevant information can and often does result in conflicting framing of a problem by different stakeholders. This is rather because of competing assumptions than because of inconsistent facts (Dunn 2004). So instead of being constituted by order and rationality, policies are more often characterised by constant paradoxes of uncertainty, interpretation, contested meaning, power, volatility, compressed views of time and space and partial information (Colebatch 2006).

Most policies are often characterised by constant paradoxes due to continuous change of external and internal developments, causing shifts in problem perception and priority setting. Often it is not clear what the real causes are and different competing policy options are on the table. Also people's understanding and interpretation change, new research results come in and new ways of using and interpreting information are used. Developments in science and technology, for example, have a strong potential to influence social change. There are, however, many reasons why the practical use of technology and scientific knowledge varies widely between countries. Societies differ, economies differ, and governments deal with international scientific developments in different ways through the policies they pursue (Timmermans 2001).

7.4.2 *The Dynamics of Policy Change*

Outcomes of a policy process are not always easy to discern at the time. Milestones when decisions are made and announced can be recorded, but their significance – they may be seen as more or less important over time – is not always clear. It can be useful to visualize this process as a series of sequential steps. In this linear model the policy process is divided into different steps: i.e. the problem definition, the analysis of alternative solutions, the adoption of a solution, and its testing and evaluation. Each step is treated as temporally and functionally distinctive. The model is most useful as a heuristic for identifying times and places where different outcomes are produced. But, the downside is that this model is oversimplifying and as such not optimal to understand policy change.

Sabatier's Advocacy Coalition Framework asserts that change in policy and policy implementation is better conceptualised as a series of interactions between groups of people in contrast with a series of transitions between stages (Sabatier and Jenkins-Smith 1993). Overtime, periods of incremental change are the norm. However, when conditions are right, a significant social change can be observed (Wood 2006). This is in line with Kingdon (1995), who suggested that realities of policy making are better captured by a focus on the flow and timing of policy action. In this model, streams of problems, solutions, and politics move independently through the policy system. Occasions arise where these three streams are joined. Policy change can be defined as an overall system behaviour that comes out of the interaction of many flows of activity. It cannot be predicted from knowledge of what each component of a system does in isolation. Net, principles of this model are:

Changes in policy and policy implementation are rarely the result from a linear process of generating research, laying out policy options, choosing between alternatives, and evaluating the implementation of the selected option. Rather, changes are the result from a process of iterative interactions between three streams of activity: defining the problem, suggesting solutions, and achieving commitment for action. Changes occur when these three streams converge, presenting a window of opportunity to effectively drive decisions.

Opposite to these change processes are a number of processes that promote policy stability. Two key processes, highlighted in the policy analysis literature, are path dependence and closed networks (Howlett and Ramesh 2002). Path dependence refers to how current decisions are influenced by the institutional and behavioural legacies of past decisions. Closed networks refer to policy stability promoted by the ability of existing key policy actors to prevent new members from entering into policy debates and discourses. Under normal policy conditions, the agenda space allocated to any issue is dominated and controlled by the formal system. Alternative visions and innovations are systematically excluded to maintain risk-averse development characterized by incremental change.

7.5 Interactions Between Strategic Intelligence Applications and Decision-Making

Inherent to the methods and theoretical assumptions used, SI applications have their advantages and limitations to support decision-making. These may be evaluated both in their own terms and in terms of whether they fit the purpose (Burton et al. 2006). But assessing the effects of SI applications requires an understanding that it is just one of the influences on public policy. To be effective it needs to be tuned into the strategic behaviour and cycles of policy and social actors (Georghiou and Keenan 2006). The given that a conventional process evaluation measures mainly activity and not its significance supports the fact that a broader perspective is needed to understand the effectiveness in the decision-making process. Three complementary perspectives on policy change are proposed to analyse the interactions between SI applications and decision-making.

The first perspective – window of opportunity- is related with the dynamic behaviour of policy change. The second perspective is related with the clarity of purpose and the third reflects on the legitimacy of policy evidence. Strictly peaking, none of these perspectives are entirely accurate in their explanation. But an analysis based on these perspectives can provide more insights on why some SI application exercises had made a difference and others not. In combination these perspectives can be used as an analytical framework to make a more in depth evaluation.

7.5.1 Window of Opportunity

Policy problems and solutions are social constructions. They are the result of a social process (Schneider and Ingram 1997). System dynamics are used in many disciplines – such as economic, social and environmental science – to describe complexity and change processes. Policy change can be seen as a dynamic, non-linear process involving a diverse range of stakeholders and giving rise to both positive and negative feedback. Complexity refers to the intrinsic relationships that arise from the interaction of agents capable in adapting to and evolving with a changing environment. As mentioned earlier, policy change can be defined as an overall system behaviour that is the result from interactions between people including different flows of activity. In this model, streams of problems, solutions and politics move independently through the policy system. Each individual flow of activity can only indirectly and incrementally impact the changes in policy and policy implementation. Changes emerge when these three streams converge; presenting a window of opportunity to effectively drive the decision (Wood 2006).

This approach emphasises the importance of barriers and incentives. Institutional structures, for example, can act as barriers in the way they promote conventional and risk-averse thinking and exclude new ideas and experiments on the political agenda. On the contrary, increased knowledge flows are important as incentives for policy change because the can act as a catalyst of change by raising awareness amongst the stakeholders and by confronting the decision-makers with new ways of thinking.

7.5.2 Clarity of Purpose

The complexity of policy choices prompts higher level of stakeholder participation. But the growing dependence of politicians to the other stakeholders can erode the trustworthiness of the politicians. SI applications affect decision-making not only by providing legitimacy to some forms of political action, but also by shaping the actors' perception of their interest as well their strategies (Dimitrakopoulos 2005). It is important for SI applications, such as foresight, to involve politicians prior to the start. The lack of commitment of politicians to SI applications, may lead to the emergence of parallel processes that can create divergence between the different flows of activity. This can eventually lead to inertia and limited opportunities for innovation.

To be effective, SI applications need a clear purpose and position in the policy process and the participants, including the decision-makers, should be aware of their role. Complete clarity concerning what decision-makers want to achieve is essential in order to design the process aiming to meet those objectives (Burt and van der Heijden 2003). It is therefore important that politicians must play an active role in the confirmation of the process design and the communication of the purpose. More ever, SI a can give policy-makers an opportunity to achieve visibility and leadership by taking the role as foresight ambassador. This approach emphasises the importance of transparency and political commitment in the SI application process.

7.5.3 Legitimacy of Evidence

Stakeholder participation and interdisciplinary research are considered to be key drivers behind improving evidence for policy (Enserink 2003). Regarding legitimacy, the point is not to judge whether an objectively correct decision have been made. The point is to explore if all key stakeholders had trust in the foresight process to provide strategic intelligence to support decision-making. Legitimacy is a question of perception. It is generally higher in cases in which policy recommendations have been produced in such a way that divergent values and beliefs of involved stakeholders and decision-makers have been respected, and opposing views and interest have been duly acknowledged (Niederberger 2005).

In Bulkeley and Mol (2003) the arguments in favour of a more participatory approach to improve the evidence base are summarised. Primarily, it helps bridging the gap between a scientifically-defined problem and the experiences, values and practices of actors who are at the root of both cause and solution of such problems. Secondly, participation helps clarifying different, often opposite, views and interests regarding the problems. Thirdly, it makes problem definitions more adequate and broadly supported. And fourthly, participation has an important learning component for the participants, being reflected in the enhanced quality of the support to decision-making.

SI applications are rooted in interdisciplinary research and they include methods of problem structuring as well as problem solving. For instance, foresight approaches recognise the need to understand the system and to identify the trends, events and weak signals that are critical. By breaking out the uncertainties, foresight can give decision-makers a view on what is actually driving the system, on the underlying structural relationships and on new emerging trends (van der Heijden 2005). This also allows for uncertainty and ambiguity in the contextual environment to be acknowledged and implications for strategy development to be considered (Burt and van der Heijden 2003). This approach emphasises the importance of knowledge integration to build robust evidence that can empower the stakeholders involved.

7.6 Case Study: A Joined Innovation Process

In this section, the three complementary perspectives were applied to analyse a foresight exercise: 'A joined innovation process: scenario workshops for rural development Flanders 2030'' The scenario workshops were organised to provide strategic intelligence in the policy process of regional rural development.

The foresight activity was linked to an intergovernmental committee (IPO) formed by the Minister President and political representatives of the regional and local governments. Objective of this committee was to have a formal setting to debate actual issues and to promote joined action. The process was linked with the phase of policy formulation and the time frame was limited due to the policy process: one year in total, from March 2005 till February 2006. The project was conducted by a consortium of governmental agencies and administration (Flemish Land Agency, SVR – Research Centre and Departments of Environment and Agriculture).

7.6.1 Window of Opportunity

Formal attempts to formulate a policy strategy for rural development did not succeed, probably due to a non productive dialogue between the different stakeholders and academic experts. Although both groups had a common interest, the differences in policy agendas on short term issues were too large to overcome. After the regional elections in 2005, rural development was high on the policy agenda. This was reflected by the allocation to the responsibilities of the Minister President.

By looking at the future, the scenario workshops provided a platform to build common understanding of the economic, social and environmental systems. The focus was more on possibilities – what if – and less on the different agenda's of the stakeholders involved. The scenarios were also an opportunity to build strategic intelligence by exploring different solutions and linking them to the problems at stake. During the process the challenges were being re-framed and most of the stakeholders were mobilised to take action. By doing so, the foresight exercise was able to negotiate around obstacles that prevent change and innovation and was able to increase the sense of urgency to support decision-making.

7.6.2 Clarity of Purpose

The formal connection with the Minister President and the intergovernmental committee provided a clear role and position in the policy process. Objectives were also disseminated in the invitation letters, during the workshops and in the project reports. This clarity of purpose facilitated the participation of key stakeholders and was essential to achieve a constructive dialogue during the workshops. In this case, the participation of the key stakeholders also provided strong recommendations concerning priority issues and policy implementation.

7.6.3 Legitimacy of Evidence

During the foresight exercise several integrated scenario's – plausible narratives about the future – were constructed. The scenarios were essential to acknowledge the uncertainty and ambiguity in the contextual environment. To enhance the legitimacy of the exercise, special attention was also paid to build trust with the decision-makers and key stakeholders involved. The objective was to achieve trust, not only on the outset of the scenario work, but also on the process of knowledge integration and linkage with the formal decision-making. For instance, additional time was foreseen during the workshops to provide information to the participants on the integration of knowledge derived from additional analyses and the participatory workshops. Gaining trust on these three levels is seen as essential for the success of the scenario workshops (Burt and van der Heijden 2003).

7.7 Conclusions

Decisions in a policy context are more complex then a single choice to be made. First of all, policy processes are, instead of being constituted by order and rationality, more often characterised by constant paradoxes of uncertainty, contested meaning, power, compressed views of time and space and so one. Secondly, changes in policy rarely result from a linear process. In contrast, changes in policy come through a process of iterative interactions among three flows of activity: defining the problem, identifying solutions, and achieving commitment for action. Changes occur when all these flows converge, presenting a window of opportunity. Thirdly, decisions must be made in a dynamic and increasingly rapidly changing environment, yet today's decisions constrain future decisions. In a nutshell, effective policy practice can be described as an interactive process where a strategic problem setting is linked to a plausible solution meeting the test of political consensus.

The core idea beyond SI applications, such as forecast, impact assessment and foresight, is that they can facilitate the policy process in a systematic and scientific way. But in practice, there is often a missing institutional link that will limit the effectiveness of the SI application (Edelenbos 2005). Three complementary perspectives on policy change are proposed to analyse the interactions between SI applications and decision-making. The first perspective – window of opportunity- is related with the dynamic behaviour of policy change. The second perspective is related with the clarity of purpose and the third perspective reflects on the legitimacy of policy evidence. Strictly speaking, none of these perspectives are entirely accurate in their explanation. But an analysis based on these three perspectives can provide insights on how to establish effective policy practice. The proposed analytical framework can also be used as a research agenda developed to improve the theoretical underpinnings of SI applications.

Acknowledgements The author is grateful to IPTS and to the Cost Action A22 network for the inspiring discussions and would like to thank all who commented on earlier versions of this chapter, with special reference to Serge, San, Riel, Ted and Ron for the support.

References

Bulkeley, H. & A.P.J. Mol, 2003. Participation and Environmental Governance: Consensus, Ambivalence and Debate. Environmental Values 12:143–154

Burt, G. & K. van der Heijden, 2003. First Steps: Towards Purposeful Activities in Scenario Thinking and Future Studies. Futures 35:1011–1026

Burton, P., R. Goodlad & J. Croft, 2006. How Would We Know what Works? Evaluation 12:294–312

Chermack, T.J. 2004. Improving decision-making with scenario planning. Futures 36:295–309

Colebatch, H.K. 2006. What work makes policy? Policy Science 39:309–321

Davies, H., S. Nutley & P. Smith (Eds.), 2000. What Works: Evidence-Based Policy and Practice in Public Services. Policy Press, Bristol

Dimitrakopoulos, D.G. 2005. Norms, Interests and Institutional Change. Political studies 53, 676–693

Dunn, W. 2004. Public Policy Analysis: An Introduction, 3rd Edition. Prentice-Hall, Englewood Cliffs, NJ

Edelenbos, J. 2005. Institutional Implications of Interactive Governance: Insights from Dutch Practice. Governance 18:111–134

Enserink, B. 2003. Creating a Scenario-Logic – Design and Application of a Repeatable Methodology. Proceedings of the 36th Hawaii International Conference on System Sciences

Georghiou, L.& M. Keenan 2006. Evaluation of national foresight activities: Assessing rationale, process and impact. Technological Forecasting & Social Change 73: 761–777

Geva-May, I. 2002. Cultural Theory: The Neglected Variable in the Craft of Policy Analysis. Journal of Comparative Policy Analysis: Research and Practice 4:243–265

Hendriks, F. 2002. Representatieve politiek in de netwerksamenleving. Problemen en perspectieven (Representative politics in the network society. Problems and perspectives) Bestuurskunde 11:266–278

Hoijer, B., R. Lidskog & Y. Uggla, 2006. Facing Dilemmas: Sense-Making and Decision-Making in Late Modernity. Futures 38:350–366

Howlett, M. & M. Ramesh 2002. The Policy Effects of Internationalisation: A Subsystem Adjustment Analysis of Policy Change. Journal of Comparative Policy Analysis: Research and Practice 4:31–50

Hughes, J.J. 2007. The Struggle for a Smarter World. Futures 39:942–954

Kingdon, J.W. 1995. Agendas, Alternatives and Public Policies, 2nd edition. Little, Brown and Company, Boston, MA

Niederberger, A.A. 2005. Science for Climate Change Policy-Making: Applying Theory to Practice to Enhance Effectiveness. Science and Public Policy 32:2–16

Nutt, P.C. 2004. Averting Decision Debacles. Technological Forecasting & Social Change 71:239–265

Rowe, G. & L. Freyer 2005. A Typology of Public Engagement Mechanisms. Science, Technology, & Human Values, 30:251–290

Sabatier, P.A., & H. Jenkins-Smith 1993. Policy Change and Learning: An Advocacy Coalition Approach. Westview Press, Boulder

Sanderson, I. 2004. Getting Evidence into Practice: Perspectives on Rationality. Evaluation 7:366–379

Schneider, A.L. & H. Ingram 1997. Policy Design for Democracy. University Press of Kansas, Lawrence, KS

Schwartz P. 1991. The Art of the Long View, Doubleday, New York

Shaxson, L. 2005. Is your Evidence Robust Enough? Questions for Policy Makers and Practitioners. Evidence & Policy 1:101–111

Tenbensel, T. 2006. Policy Knowledge for Policy Work. In Colebatch H.K. (Ed.), The Work of Policy. An International Survey. Lexington Books, Latham, MD, pp. 199–216

Timmermans, A. 2001. Arenas as Institutional Sites for Policy-Making: Patterns and Effects in Comparative Perspective. Journal of Comparative Policy Analysis: Research and Practice 3:311–337

van de Kerkhof, M. 2006. A Dialogue Approach to Enhance Learning for Sustainability. A Dutch Experiment with Two Participatory Methods in the Field of Climate Change. IAJ 6:7–34

van den Hove, S. 2007. A Rationale for Science-Policy Interfaces. Futures 39:807–826

van der Heijden, K. 2005. Scenarios: The Art of Strategic Conversation, 2nd edition. Wiley, Chichester, UK

Wood, R.S. 2006. The Dynamics of Incrementalism: Subsystems, Politics, and Public Lands. The Policy Studies Journal 34:1–16

Chapter 8
Corporate Foresight

K. Cuhls and R. Johnston

8.1 Introduction

This chapter addresses Foresight and Future-oriented Technology Analyses (FTA) in the context of their application to the world of business. While there has been a great deal of emphasis on the development and use of FTA in the public sector and by governments, less attention has been directed to the substantial growth in foresight and future analysis in business. However, on closer inspection, it has become apparent that the term Corporate Foresight (e.g., Burmeister et al. 2004) is commonly being used to describe very different things. In order to clarify the term, and identify its various uses, this paper has been structured to address separately 'FTA in business' and 'FTA for business'.

8.2 Why Foresight in Companies?

As already described in the contribution of Ron Johnston in this volume, there is already a long tradition of forecasting and later on also foresight with more emphasis on participation also in companies. The use of foresight by companies over many years, especially since the 1980s, presents a confused picture. On the one hand, Bradfield et al. (2005) suggest 'there is anecdotal evidence to the effect that scenarios declined in popularity during the 1980s' and cite Martelli's (2001) suggestion 'that while the use of scenarios comes and goes in waves, it has grown in the last one or two decades but not that much and probably less than expected'. On the other hand, the UNIDO Technology Foresight Manual (2005) reports:

> *In the last two decades several large enterprises in such diverse sectors as energy, automotive, telecommunications and information technology have established foresight groups and strategic planning processes, which analyse the long-term prospects of new technologies and their impact on markets and corporate strategies. DaimlerChrysler's Society and Technology Research Group (STRG) is one of the first future research groups to be established within a company. Since 1979 it has investigated, in close cooperation with its customers, the factors shaping tomorrow's markets, technologies and products.*

C. Cagnin et al. (eds.) *Future-Oriented Technology Analysis,*
doi:10.1007/978-3-540-68811-2, © Springer-Verlag Berlin Heidelberg 2008

A survey of 18 major European firms with substantial R&D budgets in highly competitive sectors[1] (Becker 2002) revealed that all were engaged in foresight, with a focus on technology trends, or market trends, or both. There were two principal reasons given:

> *Either they are a consequence of a companies' business operation which inherently demand such a long-term orientation (as in industries with long product cycles), or they are undertaken as a proactive step to better cope with uncertainties in the business environment in general.*

In the 1990s, as national governments became more active in foresight, large companies such as Philips, Lucent Technologies, Siemens, DaimlerChrysler and of course Shell had already developed their own systems of looking into the future and drawing conclusions to feed into their planning. They use different terms for the activities – and the methodologies differ, too. But all of them generate information for their planning purposes. The methods of choice were based on patent analysis, literature analysis, scenarios (Godet 1994, 1997; Berkhout and Hertin 2002) surveys, sometimes even in the form of a Delphi, and technology roadmaps (Reger et al. 1998). The most popular of these techniques with business were roadmaps and scenarios.

Since the 1990s, the use of foresight has spread around the world. Networks and alliances were established to facilitate this, such as EIRMA (European Industrial Research Management Association) in Europe and the Global Business Network in the US. It should also be mentioned that management consultants turned their attention to foresight in the mid-1990s. Whereas previously, the dominant management approach had been that of strategic planning, with a time horizon commonly of one to three years, writers such as Hamel and Prahalad (1994) emphasised the need for companies to play a role in inventing a longer-term future in which the organisation would have competitive advantage. Since then, the growth in the number of consulting companies offering services in foresight has been considerable[2]. Another indication is the growth in the number of conferences and workshops about 'Corporate Foresight' especially since 2005.

8.3 Types of Corporate Foresight

Corporate Foresight covers a wide range of different types and approaches to addressing the future. Projects differ in methodology, size, concepts, understanding of the time horizon and objectives, as well as the impacts they might have. In this overview paper we distinguish two major categories:

---

[1] Companies like DaimlerChrysler, Ericsson, Aventis, IBM, Philips, Siemens, BASF, Volvo, BT.

[2] A Google search of foresight produces some 22 million items (2006), with a substantial proportion from consultancies.

- Firstly, future-oriented technology analysis in business, performed directly in companies for various purposes
- Secondly, FTA for business, performed by various actors, and applied in business.

8.3.1 FTA in Business

FTA in business is conducted inside a company, in most cases by the company itself, though sometimes assisted by external facilitators or advisors. One challenge in identifying foresight in business is the variability in usage of terms: foresight and forecasting are frequently used interchangeably, or not used at all in favour of the more familiar strategic planning and management vocabulary. Some applications of foresight in business are:

1. Foresight for strategic planning

Foresight and future-oriented studies are commonly performed in companies that have their own strategic (planning) departments. These are usually larger companies such as Siemens AG, DaimlerChrysler AG, BASF AG in Germany, Lucent Technologies in the US, Nokia in Finland, or Philips in the Netherlands.

A number of broad generalisations can be drawn. First, in many cases, scenario planning approaches from the different 'schools of thought', such as Godet (1994, 1997), the US Global Business Network, the 'Nixdorf School' (Fink et al. 2001), Ringland (1998), Schwartz (1991), Van der Heijden (1997) or von Reibnitz (1988), are applied. Second, most of them integrate a form of cross-impact matrix, and examine drivers, influences, key factors, or visions of the future. Third, in most cases, alternative scenarios are developed (e.g. best versus worst case, different directions and options, but all plausible in themselves). Some work with a single normative scenario that is directly applied as a 'vision' in the business context. Fourth, the results of these activities are generally directly used for strategic vision building or planning, are performed directly in the company, and usually have a direct internal impact. As an in-house activity, the process is often as important as the results.

There are also many future-oriented projects conducted in R&D strategy departments or other divisions related to innovation. Often, these projects are not called foresight at all but labelled 'long-term strategy planning' or 'studies for long-term anticipation', 'strategische Früherkennung' or even equivalents of the word 'forecasting'. Nevertheless, it is clear that these analyses fit under what we call Corporate Foresight or FTA.

Many companies also try to gain an overview and 'to better understand the social and cultural context of the use of technology. Firms in particularly technology-intensive sectors (Philips, Ericsson, IBM, Siemens) also use foresight more broadly to build up knowledge both about emerging technologies and their future users (Becker 2002, p.9).

2. Foresight for marketing

While less common, foresight for marketing has been used to raise general societal issues by companies like Janssen Cilag (health), or Siemens AG (Horizons, Pictures of the Future). These activities are used to position the company as a 'responsible partner in society'. The impacts of these activities on the company or society are difficult to evaluate but nevertheless real.

Other Foresight activities are conducted in the marketing departments of large companies, often under different labels. Some companies look for 'social innovation' (German Telekom), trends in consumer behaviour, new patterns of consumerism (like 'event consuming') or the needs/demands of people in general. These approaches are mainly short-term and can be found in nearly every large marketing department. Internal and external data about customers are often analysed for this purpose. Some companies analyse these data further and with a longer term view, in order, for example, to develop new products. (Kondô 1993)

In 1998, we performed interviews in German, Japanese and US companies (Reger et al. 1998) about their foresight activities. In two Japanese companies, we found a system that connected the marketing and R&D departments directly in order to identify long-term weak signals on the one hand, and analyse customer behaviour, needs and claims (if something went wrong) on the other. In one of the companies, a product could be identified that was directly derived from the wishes of the customers (or better: from the description of what they disliked), and had been a huge market success not only in Japan.

3. Foresight for organisational change

In some cases, companies engage in foresight to provoke organisational change. In most cases, the aim is to restructure the internal organisation in order to position for possible events to come. However, as in so many cases of organisational change management, internal resistance to change is so strong that there is limited impact. An example is Deutsche Telekom (Reger et al. 1998) which aimed to install a foresight system but concluded that there was already a significant level of foresight being conducted under various names.

Organisational change needs a lot of internal knowledge and the persons involved in performing such a foresight process often need to work locally in the company. Without being integrated for a certain time in the organisation and without building up 'trust', the impact can be very limited. This is for example the philosophy of DaimlerChrysler's future group (Society and Technology Research Group) in Berlin.

4. Foresight for innovation

Foresight for innovation does not only deal with technologies but also their applications and potential markets for new products derived from these new applications. The methods do not differ from other corporate foresight approaches on a smaller scale but are often rather short-term oriented as the companies have an interest in being on the market with a profitable product as soon as possible.

Others have relatively long-term view, even in companies 10–20 years ahead. Like in national foresight approaches, here often the different perspectives or the framework conditions are in the forefront to be able to go a step further in innovation (Ruff 2007; Schneider 2007). As an example, the development of the Smart automobile, DaimlerChrysler, now Daimler AG, is mentioned. Deutsche Bank AG develops specific maps to identify for example the driving forces for structural change (Hofmann et al. 2007) in order to anticipate changes that might have an effect on their products at an early stage.

In innovation foresight, there are attempts to establish methods beyond the classical repertoire as it is very difficult to assess a future "market". A number of larger companies, such as Hewlett-Packard, Intel and Google have pioneered the use of 'prediction markets' as a way of tapping and applying the knowledge and experience of all their staff in making judgements, through a virtual trading mechanism, about likely future directions of technology development. (see for example The Foresight Exchange Prediction Market, Pennock et al. 2001, see also an experiment with prediction markets in Europe concerning the Creative Content Industries, Luckner and Weinhardt 2007; Spann and Skiera 2003; Wolfers and Zitzewitz 2004; Ortner 1997; Gräfe et al. 2007; Gräfe and Cuhls 2008). With the current technical means to support such analyses, this is a rather new development and only few impacts can be observed until now.

Other companies perform innovation foresight as a prerequisite for their own strategic planning (Philips). Some firms regard the catalytic function (see below), to stimulate and enhance their innovation processes, as important. Becker (2002, p.10) mentions DaimlerChrysler, Philips and Decthlon in this context.

8.3.2 *Foresight for Business*

This second category involves the application of the results of foresight activities performed outside a particular business. Here, often, the activities are not directly tailor-made to the objectives of the companies but they are nevertheless used internally or for business purposes. These kinds of activities include:

1. Making use of the results of national and other public foresight activities

One of the most popular approaches is using data (e.g. surveys, Delphi results) from national foresight activities for business and especially for strategic planning purposes. Japanese companies have reported effective use of the results of the regular NISTEP Delphi reports. Similar findings have been made for German industry (Cuhls et al. 2002). Only in Japan, a series of foresight activities can be found that provide data directly, and on a regular basis, for external (business) use. In all other countries, the public foresight projects or programmes do not run in such a continuous way or are performed with a more process – rather than output-oriented methodology (like in the German Futur, the UK Foresight etc.).

But as the data are more general in nature, they have to be adapted to the different users and purposes. For sectoral analyses, the relevant data have to be searched, selected, assessed, and then have to be qualitatively transferred into the context. Even for gaining an overview, they have to be clustered or processed in a way that the overview is adapted to the context and objectives. Otherwise, no impact can be achieved.

2. Foresight by industry associations

Industry associations have conducted foresight exercises or analysed the results of national foresight activities for their members; e.g. the German ZVEI (electro technical association) and the VDMA (machine tools). Currently, the VDMA is establishing a 'manufacturing platform' to provide information for their members. An influence from the international project ManVis (www.manufacturing-visions.org) can be observed in this case.

3. Foresight by foundations

Foresight by Foundations is normally targeted at setting priorities or providing information for society (e.g. European Science Foundation) or for companies, particularly small and medium-size enterprises (SMEs). The German Stiftung Industrieforschung, which provides money for research institutions that work with SMEs, performed two foresight projects to identify interesting thematic fields. In two survey rounds, new fields and research topic were identified and then assessed according to categories of importance and if the companies would invest in the topic themselves. Based on the survey results, 'interesting' themes were selected. In these fields, interviews were performed to gather detailed information about the research questions that are relevant for the future and fit into the funding portfolio of the Foundation. From the results, ten topics were selected for support in the following year. The topics were e.g. laser diode systems and medicine technologies. The impact here was in the setting of priorities, a direct call for applications, and in the consequential research results (for details see Cuhls 2007).

4. Multi-client studies

There is also potential for foresight as a multi-client study. These studies are often financed by the companies themselves or by e.g. a ministry, the European Commission or an association (see above) to promote future developments. An example is 'HyWays', an international consortium to promote hydrogen infrastructure. In the project, a validated and well-accepted roadmap for the introduction of hydrogen in the energy system is being developed. Companies like Air Liquide, the BMW Group, Det Norske Veritas, DaimlerChrysler AG, Total, GM Opel, Vattenfall Europe and others are participating as well as a range of institutes.[3] The Member

[3] French Atomic Energy Commission, Energy Research Centre of the Netherlands, Italian National Agency for New Technologies, Energy and Environment, Imperial College of Science, Technology and Medicine, Fraunhofer ISI, Instituto de Engenharia Mecanica, L-B-Systemtechnik, Université Louis Pasteur and the Zentrum für Europäische Wirtschaftsforschung.

State Representatives are from the French Atomic Energy Commission, the Italian National Agency for New Technologies, Energy and Environment, the German Energy Agency, the Hellenic Institute of Transport, Senter Novem and the Western Norway Research Institute.[4] In these cases, there are direct impacts on companies because the objectives and methods of the studies are tailored to their needs. On the other hand, this is only possible, if the questions asked are relatively clear and if the need for information is recognised in the specific field.

8.4 Objectives and Methods of FTA in and for Business

Five major objectives of corporate FTA have been identified (Becker 2002, p.9):

- Anticipatory intelligence, i.e. providing background information and an early warning of recent developments
- Direction setting, i.e. establishing broad guidelines for the corporate strategy
- Priority setting: i.e. identifying the most desirable lines of R & D as a direct input into specific (funding) decisions
- Strategy formulation, i.e. participating in the formulation and implementation of strategic decisions
- Innovation catalysing, i.e. stimulating and supporting innovation processes between the different partners

However, sometimes the aims are even broader (Cuhls 1998; Burmeister et al. 2004). The impacts of the processes of FTA studies are accordingly different.

The way the foresight or FTA function is organised in companies, also varies a great deal. Becker (2002) identifies three approaches: 'The Collecting Post' with limited but integrated capacities which provides background information for decision-making processes, 'The Observatory' is an autonomous foresight unit, with staff and a budget of its own focussed on highly specific company objectives, and 'The Think Tank', which operates fairly independently across a wide range of areas. Applying this classification and drawing on literature analysis and our own interviews, we have classified a number of international companies according to their relative focus on external information gathering versus internal change (Fig. 8.1). In reality, some companies have different things, e.g. observation posts as well as think tanks.

In the above mentioned survey for Deutsche Telekom in 1998, in which US, Japanese and German companies with foresight activities were identified and interviewed, it became obvious that the terminology for foresight and FTA is diverse but the methodologies used are often the same. The following methods were identified (Reger et al. 1998, Reger 2001; see also Burmeister et al. 2004, p.37 and Schwartz 2006):

[4] More details can be found at: www.hyways.de

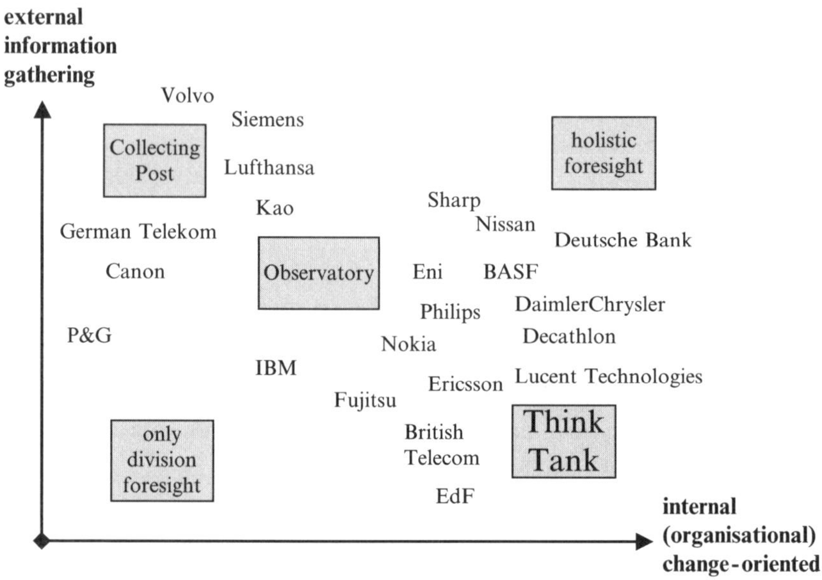

Fig. 8.1 Internal versus External Orientation of Business Foresight

- Patent analysis
- Publication analysis
- Market analysis, environment analysis, trend studies
- Extrapolations
- Competitive reports and benchmarking analyses
- Competitive intelligence
- Systematic customer surveys
- 'Intelligent' cost analysis
- Co-wording/ co-heading analysis
- Risk analysis
- Simulations
- Life cycle analysis
- Internet search systems and engines
- Technology mapping
- Expert surveys/ Delphi surveys
- Relevance tree analysis
- Portfolio analysis
- Scenarios
- Future labs
- Future workshops
- Future conferences
- Technology or product roadmaps
- Creativity methods (brainstorming, meta-plan, mindmaps)

Various authors (e.g. Burmeister et al. 2004, p.37; Schwartz 2006; Cuhls and Kuwahara 1994; Reger et al. 1998) have indicated that trend and environment analyses, creativity methods, scenario methods and expert surveys (including Delphi surveys) are the most commonly used tools for FTA in firms. Trend extrapolations are also applied but must be carefully analysed, to identify when the trend might break. In companies, the classical and rather quickly performed analyses of publications, of product life-cycles and of business portfolios are still the standard whereas more time- and resource-intensive methods are less common. Although some companies prefer the long-term view, the most commonly found approaches are short-term.

In the US and Japan 10 years ago (Reger et al. 1998), it was evident that roadmaps were much more recognised as an FTA or foresight tool than in Germany (and other European countries). There is evidence that this has substantially changed, with a number of roadmapping projects completed or in progress. Nevertheless, the term means different approaches which all result in a timeline format for listing science and technology in their contexts (Weissenberger-Eibl and Speith 2006 and 2007; Albright and Kappel 2003; McMillan 2003; Phaal 2002; Phaal et al. 2001; Kostoff and Schaller 2001; EIRMA 1997, Barker and Smith 1995; Willyard and McClees 1987).

A Delphi survey by Schwartz (2006) about the application of FTA or foresight methods in corporations points to an increase in applications. This can also be interpreted as a sign that FTA is being more directly integrated in companies.

If scenario methods in their different variants are applied, ‘wild cards’ (Steinmüller and Steinmüller 2003) are commonly considered to prepare for unforeseeable events, ‘think the unthinkable’, and prepare for a worst possible case. This is an addition to many scenario processes and generates more sophisticated surprise-connected scenarios as well as policy recommendations.

The different methods have their different objectives, applications and therefore impacts. At this stage, it is premature to attempt to identify best and worst practices. What is clear is that the methodology has, as in all FTA exercises, to be specifically tailored to the objectives of the company, and the future issues of interest.

8.5 Impacts of Corporate FTA and Outlook

It can be concluded that there has been a very substantial growth in the application of FTA in the business sector, with a range of diversified impacts. In some cases, direct consequences flow from the foresight: e.g. new business areas, new products, new production procedures, new strategies and targets, and new forms of organisation.

The different methods applied in business and for business lead to different applications and impacts. But the consequence is that it is very difficult to trace the different results back to the different outputs of FTA studies because there are so many intervening factors involved. Process-orientation, as an essential part of FTA activity can also be an integral and important aspect, because this has a direct impact on the people in business, their culture, and their behaviour. This is not only true for the decision-makers but all persons involved in the process.

There remain many opportunities for the more extended application of FTA in the business sector. In the traditional area of strategic planning, FTA approaches have much to offer in the development of visions and goals, in the analysis of the environment external to the organisation and the opportunities and threats it may pose in the future, and in the development of strategic intelligence. In risk management, there are enormous opportunities for the better identification of potential future risks, and of the stakeholders who may be affected. Even in change management, as we have noted, developing and communicating a clear image of the future and using the group processes of FTA to motivate changes in behaviour offer considerable potential.

Becker (2002) identified some additional points in corporate foresight approaches that need strong improvement: From the methodological point of view, foresight sometimes also needs a better grounding regarding quantitative analyses and economical modelling, in order to achieve a better quality and even greater accuracy of its results. Often, the results that are generated are not adapted to the target groups. Sometimes a better commitment can be achieved if these groups are better integrated. For this, the language of the target group has to be applied, which sometimes needs a kind of translation work.

As we have seen, there are more centralised and more de-centralised approaches in foresight. But some of them are too fragmented and specialised to give a complete picture, others are too aggregated to be precise enough for decision-making. To find the right balance is not an easy task. And the way foresight is organised needs an additional adaptation to the corporate culture. Without a kind of "spirit" for being open for long-term thinking, foresight looses ground very soon in an industrial surrounding. In larger companies, a kind of "shareholder value mentality" hampers this long-term thinking even more.

The major difficulty remains that of demonstrating unequivocally, and preferably quantitatively, the impact of an FTA study in or for a company. In many cases, even the feedback of the users is missing. The process benefits in certain cases are clearly demonstrable, but often fail to persuade more 'bottom line' focused companies. In business, everything has to be judged on the basis of money or other added-value that can be quantitatively measured. Some companies already have a system for this. But these are mainly larger companies which have already acknowledged that FTA might be an add-on and that the company itself might profit from it even if the quantification is weak. Here, there may be opportunities for learning from the experience of foresight evaluation. However quantifying the precise benefits will probably remain a long-term goal.

What is clear in corporate foresight: Doing foresight for its own sake does not make sense but clear goals have to be se in order to be achievable.

References

Albright, R.E. and Kappel, T.A. (2003) Roadmapping in the Corporation, Research Technology Management, 42(2), pp. 31–40

Barker, D. and Smith, D.J.H. (1995) Technology Foresight Using Roadmaps, Long Range Planning, 28(2), pp. 21–28

Becker, P. (2002) Corporate Foresight in Europe: A First Overview, Working Paper, European Commission, available at ftp://ftp.cordis.lu/pub/foresight/doc/st_corporate_foresight_040109.pdf

Berkhout, F. and Hertin, J. (2002) Foresight Futures Scenarios: Developing and Applying Participative Strategic Planning Tool, GMI newsletter, Available online at ttp://www.greenleaf-publishing.com/pdfs/gmi37ber.pdf

Bradfield, R., Wright, G., Burt, G., Cairns, G., and van der Heiden, K. (2005) The Origins and Evolution of Scenario Techniques in Long Range Business Planning, Futures, 37(8), pp. 795–812

Burmeister, K., Neef, A., Beyers, B. (2004) Corporate Foresight, Unternehmen gestalten Zukunft, Murmann Publishers, Hamburg

Cuhls, K. (1998) Technikvorausschau in Japan. Ein Rückblick auf 30 Jahre Delphi-Expertenbefragungen, Physica, Heidelberg

Cuhls, K. (2007) Foresight: 10 Topics for an Industrial Foundation, International Journal of Foresight and Innovation Policy, Special Issue on "Impact of Foresight Studies on National Innovation Capability", 3(4), pp. 435–446

Cuhls, K. and Kuwahara, T. (1994) Outlook for Japanese and German Future Technology, Comparing Technology Forecast Surveys, Physica, Heidelberg

Cuhls, K., Blind, K., and Grupp, H. (2002) Innovations for our Future. Delphi '98: New Foresight on Science and Technology. Technology, Innovation and Policy, Series of the Fraunhofer Institute for Systems and Innovation Research ISI no. 13, Physica, Heidelberg

EIRMA (1997) Technology Roadmapping – Delivering Business Vision, Working Group Report, European Industrial Research Management Association, Paris, No. 52

Fink, A., Schlake, O., Siebe, A. (2001) Erfolg durch Szenario-Management. Prinzip und Werkzeuge der strategischen Vorausschau [Success Through Scenario Management. Principles and Instruments of Strategic Foresight]. Campus, Frankfurt/ New York

Godet, M. (1994) From Anticipation to Action. A Handbook of Strategic Prospective, UNESCO France

Godet, M. (1997) Scenarios and Strategies. A Toolbox for Problem Solving, Cahiers du LIPS, Special Issue: Paris

Gräfe and Cuhls (2008) Prediction Markets vs. Real-time Delphi: An Experiment in the Creative Content Industries (forthcoming)

Gräfe, A. Luckner S. and Weinhardt C. (2007) Prediction Markets – A Chance for Process-Based and Socially-Embedded Organizational Foresight, 3rd International Conference on Organisational Foresight

Hamel, G., and Prahalad, C. (1994) Competing for the Future, Harvard Business School Press, Boston

Hofmann, Rollwagen, I., and Schneider, St. (2007): Deutschland im Jahr 2020 – Neue Herausforderungen für ein Land auf Expedition, Aktuelle Themen 382, Deutsche Bank Research, www. expeditiondeutschland. de, access 10/10/2007

Kondô, S. (1993) R&D Senryaku Ritsuan no tame no 'Gijutsu Yosoku' Katsuyô Gaidobukku (Handbook on technology foresight activities for the strategic R&D planning), Tôkyô

Kostoff, R.N. and Schaller, R.R. (2001) Science and Technology Roadmaps, IEEE Transactions of Engineering Management, 48(2), pp. 132–143

Luckner, S. and Weinhardt, C. (2007) How to Pay Traders in Information Markets? Results from a Field Experiment, Journal of Prediction Markets, 1, pp. 1–10

Martelli, A. (Summer 2001) Scenario Building and Scenario Planning: State of the Art and Prospects for Evolution, Futures Research Quarterly

McMillan, A. (2003) Roadmapping – Agent of Change, Research Technology Management, 42(2), pp. 40–47
Ortner, G. (1997) Forecasting Markets – An Industrial Application: Part I, TU Vienna, Working Paper, http://www.ibab.tuwien.ac.at/apsm/fmaia2.pdf
Pennock, D.M., Lawrence, S., Giles, C.L. and Nielsen, F.A. (2001) The Power of Play: Efficiency and Forecast Accuracy of Web Market Games. In: Technical Report 2000-168, NEC Research Institute
Phaal, R. (2002) Foresight Vehicle Technology Roadmap – Technology and Research Directions for Future Road Vehicles, UK Department of Trade and Industry, URN 02/933
Phaal, R., Farrukh, C.J.P. and Probert, D.R. (2001) T-Plan – The Fast-Start to Technology Roadmapping: planning your Route to Success, Institute for Manufacturing, University of Cambridge, Cambridge
Reger, G., (2001) Technology Foresight in Companies: From an Indicator to a Network and Process Perspective, Technology Analysis & Strategic Management, 13, pp. 533–553
Reger, G., Blind, K., Cuhls, K. and Kolo, C. (1998) Technology Foresight in Enterprises. Main Results of an International Study by the Fraunhofer Institute for Systems and Innovation Research (ISI) and the Department of R&D Management, University of Stuttgart, Karlsruhe/Stuttgart
Reibnitz U. von (1988) Scenario Techniques, McGraw-Hill Book Company GmbH, Hamburg
Ringland, G. (1998) Scenario Planning: Managing for the Future, Wiley, Chichester
Ruff, F. (2007) Corporate Foresight – Shaping New Realities in Urban Mobility: A Case Study of Foresight-Driven Innovation in the Automotive Industry, (http://tf-wpii.cybertest.cz/dokums_pres/tf_panels_ruff_7.pdf, access 18/10/2007)
Schneider, St. (2007) The Future of Foresight in Innovation and Competitiveness, UNIDO Expert Group Meeting of Future of Technology Foresight, Vienna, May 29th and 30th, 2007 (http://www.unido.org/file-storage/download/?file_id=67610, access 16/10/2007)
Schwartz, P. (1991) The Art of the Long View: Planning for the Future in an Uncertain World, Currency and Doubleday, New York
Schwartz, J. (2006) Die Zukunft der Zukunftsforschung im Deutschen Management: eine Delphi Studie (The future of future's research in German management: a Delphi survey), Executive Summary, part of a dissertation, Universität der Künste, Berlin
Spann, M. and Skiera, B. (2003), Internet-Based Virtual Stock Markets for Business Forecasting, Management Science, 49, (2003) 1310–1326
Steinmüller A. and Steinmüller, K. (2003) Ungezähmte Zukunft. Wild Cards und die Grenzen der Berechenbarkeit (Untamed Future. Wild Cards and the limits of predictability), Gerling Akademie Verlag
UNIDO (2005), Technology Foresight Manual, Vienna, available at http://www.unido.org/data/forms.cfm?formcode=TF-MANUAL-REG
Van der Heijden, K. (1997) Scenarios. The Art of Strategic Conversation, Wiley, Chichester
Weissenberger-Eibl, M., Speith, S. (2006), Flexibles Roadmapping – Eine Methode für die Vorausschau und Technologieplanung im Umfeld technologischer Durchbrüche, in: J. Gausemeier (ed) Vorausschau und Technologieplanung. HNI-Verlagsschriftenreihe. Paderborn, vol. 198, Heinz Nixdorf Institut, Universität Paderborn, pp. 396–424
Weissenberger-Eibl, M., Speith, S. (2007) Disruptive Technology Forecasting in Companies – Framework and Application, Presentation at the 16th IAMOT Conference. Miami Beach, FL
Willyard, C.H. and McClees, C.W. (1987) Motorola's Technology Roadmap Process, Research Management, September–October, pp. 13–19
Wolfers, J. and Zitzewitz, E. (2004) Prediction Markets, Journal of Economic Perspectives, 18, (2004), pp. 107–126

Chapter 9
The Higher Education Sector and its Role in Research: Status and Impact of International Future-Oriented Technology Analysis

L. Georghiou and J.C. Harper

9.1 Introduction

In recent years, the higher education sector has increasingly been perceived as a key part of innovation systems at all levels of analysis, including national and regional, and through the eco-system which links large and small firms together and with their collaborators (Coombs and Georghiou 2002). The core functions of Universities, training and basic research, have been subject to external forces, some of which have already made their effects felt, while others are keenly debated as societal expectations of the sector change. These activities have been supplemented by a drive towards the Third Mission, relating Higher Education Institutions (HEIs) to their socio-economic and cultural context. As with, what are in many cases long-standing institutions which are either in the public sector or rely heavily upon its funding, the sector has also felt the pressures of public sector reform in its managerial and accountability structures. Despite an experience of major changes such as massification of student access, technological change, funding models, specialisation of mission, growth of research activity and internationalisation in all respects, there is a continuing expectation that further changes are coming and hence an apparent need for Future-Oriented Technology Analysis (FTA) activity to help institutions and their stakeholders to go forward.

There is a growing politicisation of the Higher Education (HE) sector manifested in ongoing reform processes at different levels, for example in the EU Member States, where Lambert and Butler (2006) have summarised the challenges faced. In Japan the transformation of National Universities to "independent administrative institutions" (agency status) has been accompanied by a wide range of further restructuring and reform. As an example of a comprehensive national review the work of the United Kingdom's National Committee of Inquiry into Higher Education (Dearing Report 1997) could be cited. Commissioned to advise on the development of Higher Education on a 20-year time horizon this implicitly involved development of a future vision but also resolved a particular political problem, that of how to introduce student fees into a system that had previously been paid for entirely by government. Change has also proceeded at regional and university levels. The range of stakeholders engaged in the international (worldwide) debate over the

C. Cagnin et al. (eds.) *Future-Oriented Technology Analysis,*
doi:10.1007/978-3-540-68811-2,

future of Higher Education is reflected in the variety of FTA activity described in Sect. 2 of this paper.

To be clear on the scope of this chapter, it is important to emphasise that the focus here is not the role of higher education as a vehicle for the development and execution of FTA studies. This is of course important and has itself been the subject of study (Slaughter 1998). However, here we examine higher education as an *object* of FTA. This extends to the institutional and sectoral use of FTA in HE and the impacts generated on policy and decision-making. While our interest, as stated above, is founded in the future role of universities in the research and innovation system, the integrated nature of universities makes it unwise to divorce completely this aspect from their educational role. Hence, we are concerned with FTA as applied to the composition of the higher education sector, the role and nature of institutions, and methods and practices in research and training. The research dimension provides a particular link to the broader concerns of FTA. For most developed countries without a communist history, universities constitute the major performers of basic research; in some, such as France, Germany and Spain, their relation with national research organisations is also an important part of the future picture.

We have been restricted by the limited instances of FTA activity in HE or more accurately by the available documentation of it. We are often limited to selecting from the much larger literature on higher education in some studies which at least engage with the future explicitly though they may not use what are commonly recognised as FTA approaches. See for example Etzkowitz et al. (2000) who analyse trends to chart the emergence of the entrepreneurial university, but also note that:

> *"Firms, universities and governments who, individually and collectively, engage in 'bottom up' planning, 'roadmapping' and foresight exercises are more likely to reap future rewards than their peers focused on the short-term."*(p.327)

Despite these constraints, we aim to capture some of the commonalities in content of the FTA work that does exist and where possible to analyse approaches, particularly in studies which emphasise the role of HEIs in knowledge production. Key contrasting elements of international FTA in HE and national and university-based FTA-type exercises, in terms of stakeholder involvement, approaches and content, are analysed together with the results generated. Finally we briefly discuss the realised and potential impact of FTA for this sector and highlight some of the key challenges ahead.

9.2 The Landscape of FTA on Higher Education and its Role in Research

The future of higher education and in particular of universities has been the subject of extensive discussion and study but little of this has been in the context of explicit FTA methods. Extant work is distinguished more by content than by process. Reviewing the literature, Skolnik (1998) noted the difficulties of identifying what higher education would look like in the twenty-first century and concluded that future scenarios are presented without much evidential or analytical basis, making it "difficult to distinguish forecast from prescription or wish". This effect may have been exaggerated by the pre-millennial environment from which the reviewed materials emerged. Nonetheless, some significant works and activities in the 1990s may be noted, for example in the first UK Foresight Programme where the Leisure and Learning Panel addressed this issue (followed up in the second programme by a collected edition of commissioned and extant essays on universities in the future (Thorne 1999). Much of this work originated in academic work in the higher education studies community. Also in this category, more recently, the Centre for Higher Education Policy Studies (CHEPS) at Twente University carried out a Delphi study "European Higher Education and Research in 2020) subsequently used to support scenarios (Kaiser et al. 2004).

International organisations have been particularly active in producing such studies, the significance of which we will return to in the concluding discussion. The EU DG Research Foresight Unit set up two expert groups on HE and Research in 2001 and 2002. The first identified possible scenarios for 2015, highlight major trends and challenges, including demographics, student consumerism, diversification and differentiation of agents and functions, as well as pressure for accountability and impact on governance (Strata-ETAN 2002). The second group was in effect an effort to link the first to policy impact and focused on two objectives, strengthening the Higher Education Research system itself and enhancing the system's relations with its environment, by awareness activities, regulation and open coordination, new research areas and new actions on HE. A third group is currently looking at the future of "key actors" in the European Research Area, including universities, along with industry and others. Results of this last exercise were recently summarised by Havas (2008), showing "cascading visions" through three levels: overall EU policies in a global context, the future of the European Research and Innovation Area, and that of individual universities.

The OECD has also been an active contributor through its University Futures project which is designed to inform and facilitate strategic change to be made by government decision-makers and other key stakeholders in higher education. Scenarios are more driven by educational changes, but with some research visions have been produced by the OECD Centre for Educational Research and Innovation (OECD/CERI 2005). Six scenarios are mapped across two key dimensions, the range of recognised educational supply and the range of educational participation (Vincent-Lancrin 2004). More recently, a set of four scenarios from the same origin have focused on academic research, locating it in a possibility space with two dimensions: administrative versus market forces; and international focus versus national focus (Vincent-Lancrin 2006). These are discussed further in Sect. 9.4. There is evidence of growing partnerships between the World Bank, UN agencies and OECD on key policy concerns related to human development with a view to promoting joint approaches and efforts. In the area of higher education, OECD and the World Bank are exploring "whether and under which conditions cross-border higher education could benefit the developing countries' capacity building agenda".[1]

[1] See http://www.oecd.org/document/28/0,2340,en_2649_35845581_37188956_1_1_1_1,00.html

The UNU Millennium Project did not directly address Higher Education per se, but in its scenarios on Future Science and Technology Management Policy Issues it raised several issues in the domain of the moral and political status of science that could be seen as relevant to this area (Glenn and Gordon 2004). However, an important output from the Millennium Project stressed the importance of HE in development and called for a reshaping of universities involving "adjustments in curricula, changes in schemes of service, modifications in pedagogy, shifts in the location of universities, and the creation of a wider institutional ecology that includes other parts of the development process." (Juma and Yee-Cheeong 2005).

Similarly UNESCO's prospective work has focused on key emerging development and equity concerns in the HE sector within the broader context of the rise of knowledge societies. In its 2005 World Report on Knowledge Societies it warns of the risks linked to the move towards a global market in higher education. "The risks of "commoditization" in the field of higher educational are very real even if all countries do not find themselves in the same situation in relation to such challenges. Those with a long university tradition are generally less threatened by this diversification of higher educational provision. The most worrying case is that of countries lacking a university tradition: the advent of knowledge societies is often linked to the emergence of full-scale *markets* in higher education. This has prompted some commentators to speak of the "Macdonaldization" of knowledge. There is a need to ensure that these trends do not lead to a distortion of the original missions of higher education".[2] The Report highlights the fact that whilst the more advanced countries lead by the US have invested in proactive policies aimed at commodifying and internationalising the HE sector, developing countries are relegated to the role of consumers in the global market for HE services. Other key concerns relate to the strategic interest of high performance HE institutions as players in their own right in international competition and the growing concentration of resources in world class centres of excellence. The more future-oriented part of the report focuses on why the future university does not exist and reflects on the challenges to major reforms to university curricula to reflect more holistic approaches and content. The Report concludes by highlighting on the one hand the potential for HE to provide a springboard for developing countries but also identifies the barriers related to access to knowledge (digital and other divides) and resources in general.

The activities described above clearly indicate the entry of international institutions into the higher education policy domain. Nonetheless, in many countries the national level remains the key focus for policymaking with national governments controlling both funding and the legislative environment. In others, particularly in Federal States, regional government has been the most important level and in some cases has increased its role and responsibility. One might say the normal situation is a fluid multi-actor and multi-level of governance. From an FTA perspective, this means that the location of its stakeholders is likely to affect the nature of work carried out.

[2] See http://unesdoc.unesco.org/images/0014/001418/141843e.pdf Chap. 5 p.87.

An example of a national exercise which has sought to engage a wider range of stakeholders is the current work of IVA (the Royal Swedish Academy of Engineering) which is running a project on "The University of the Future" (Heinegard 2005). Reflecting the Academy's mission this has a bias towards scientific and technical subjects but does address the full range of HE with a 10–20 year time horizon. Participants include the Swedish National Agency for Higher Education, SUHF (the Association of Swedish Higher Education), research institutes and the Federation of Swedish Industries, plus project financiers and students. The purpose of this project is to support Swedish universities so they can develop into central institutions in the growing, global knowledge economy. The approach is through three expert panels: Funding instruments and sources, Organization and specialization, and Mobility, qualifications and recruitment. These have consulted through roadshows and their works forms the basis of a synthesis report.

Stakeholding in FTA can also be located within the internal governance of Higher Education. In a few cases this is collective, as for example with the response of the Australian Vice-Chancellors Committee (the council of University Presidents) to a government review of the HE sector (AVCC 2002). A series of plenary sessions of that organization developed a vision for universities in 2020, an overview of the direction in which they considered that university funding and regulation needed to move to achieve that vision, and set out their view of the working principles against which all proposals for change should be assessed. The vision was encapsulated mainly in terms of a rise in global rankings and the role of the sector in the economy, and reflecting the political role of such visions the main emphasis of the report was upon the present need for investment and reform.

Collective visions, either from the sector or in the government context are typically constructed in an interactive way with the visions of individual HE institutions. Reviewing the presence of these on websites most typically one can see rather generic mission statements. Much less common are visions which are genuinely being used to drive change and are the product of collegial processes within the institution, hence meeting the basic criteria for what could constitute FTA activity. In this latter category we could note that the University of Melbourne operates a process to produce its strategy, calling both process and outcome *Growing Esteem*. Led by the governing body, staff and students have opportunities to participate through task forces and consultative processes. The driver is the need to develop a new funding model by 2015 based on the Triple Helix metaphor.[3]

The University of Manchester's 2015 Vision characterises that institution's post-merger drive to raise itself into the world elite and is made concrete by a set of targets expressed as key performance indicators (University of Manchester 2006). The University of Manchester was also involved in a regional foresight exercise focussed upon business-industry links in the Manchester City-Region, with a particular emphasis upon the role of Manchester Science Park (Cassingena and Georghiou

---

[3] See http://growingesteem.unimelb.edu.au/strategicplan/vision.html

2005). In Ireland DCU is currently undertaking a foresight exercise using expert groups with internal and external membership to identify strategic priorities for research. This is embedded in a 3-year cycle of strategic planning and linked to external assessment.

9.3 Methods, Approaches and Participation

In all of the studies mentioned above only three broad approaches to looking at the future can be seen:

- Extrapolation of perceived current trends
- Expert opinion, almost always in the context of an individual or panel view rather than wider consultation
- Assembly of the above into scenarios

Time horizons range from 10-20 years with little or no activity outside these boundaries. One might ask why the range of FTA tools used has been limited to this narrow range of activity, with for example a low to limited presence of consultative or participative approaches compared with other foresight exercises, even in the research domain. Our observation from reviewing the stakeholder composition of these exercises is that at one end international exercises tend to involve solely academics, however some university-driven strategy processes have sought to engage more tangibly with the broader socio-economic context by bringing local players including regional and city authorities, local developers and policy-makers together with the business community. Students and the public at large are normally distinguished by their absence. In almost all FTA activities, and especially in technology foresight, a large proportion of participants come from the academic community (Cuhls and Georghiou 2004; Klusacek 2004; Nedeva et al. 1996). This applies in part to the operators of such exercises and much more so to the experts who contribute. It is perhaps not surprising then that when FTA turns to the very domain in which such experts live their daily lives, they are not inclined to look outside. The result is a prevailing view or mindset that foresight on HE is best left to the experts. In turn, this limited concern with the broad range of stakeholders could influence the range of methods which are considered and eventually used in these exercises. HE FTA activities are this somewhat out of touch with "Third Generation Foresight" and its involvement of social and industrial stakeholders (Georghiou 2001).[4] It could

[4] The generational model proposes a series of ideal types in which first generation foresight was forecasting carried out by experts often with a technological structure, second generation worked on the science-industry nexus with participation from both groups and a sectoral structure, while third generation extended to social stakeholders and a more thematic structure. Fourth and fifth generations have been proposed subsequently emphasising distributed models.

also be posited that the role played by an academic in a specialist context is different (perhaps more disinterested and objective) from that played when acting as a member of the academic profession.

9.4 Content of FTA on HE: Common Drivers of Change

At the sectoral level a number of commonalities may be observed. If we examine the drivers of change identified several occur in all studies. These are summarised in Fig. 9.1:

Globalisation is seen as bringing competition for the right to provide training, award degrees and differentiation within the sector (Thorne 1999). The European Forum on University-based Research presents this as a strategic challenge for European institutes to balance cooperation against competition in a landscape where India and China represent new sources and destinations for the best researchers. Also reporting to the Commission, the Strata-Etan Expert Group (2002) saw globalisation as eroding the function of national education systems as central agents for national integration, again leading to greater competition and problems of social cohesion. A reduced role for the State is also a driver of marketisation of knowledge and the rise of neo-liberal ideology and associated management practices in universities. Beyond the Eurocentric view, a transatlantic dialogue on the future of higher education also identified globalisation as part of an "unholy trinity" driving change (Green et al. 2002). Some aspects created opportunities – the prevalence of the

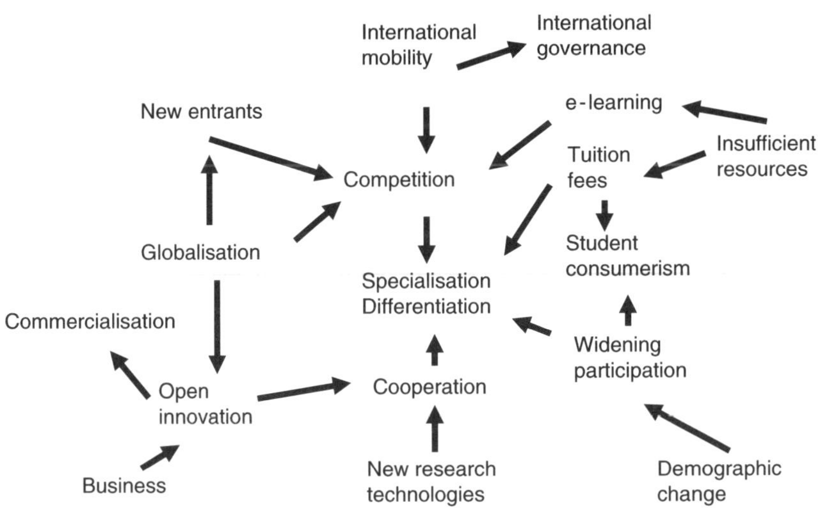

Fig. 9.1 Drivers of change for higher education

English language rather than threatening "small language" countries actually opens up international markets as they increasingly offer courses in that language. An imperative for institutions to internationalise raises the issues mentioned in the other studies and also presents a challenge to prepare students for the global competence they will require in the knowledge society.

Competition and student consumerism represent a related driver of change. The Transatlantic study saw student demand driving competition on both sides of the Atlantic with students seeking more flexible programmes, better teaching and more flexible institutions, while institutions compete for the most able students. The Bologna process and the increasing incidence of fees create the comparability necessary for a market to operate and also the entry of new players including the corporate sector itself. Research is inherently competitive but its rising costs mean more concentration in leading institutions, initially nationally and perhaps internationally.

Rise of new agents and functions, extending to a variety of intellectual entrepreneurs, including virtual Mega-universities, corporate universities and academic brokers are entering the markets traditionally occupied by universities (Abeles 1998; Kirp 2003).

How did they emerge? How are they affecting the dynamics of HE/R systems?

Are they a passing phase or will they become stronger features in the system? What are the negative and positive impacts? What can governments do to regulate their operation?

Demography provides its own pressures. The Strata-Etan study noting Europe's ageing populations concludes that adaptation to cater for the needs of older students is a necessity exacerbated by potential effects of retirements among Faculty. In the meantime increasing participation, often referred to as the massification of education, though now well-established as a phenomenon, has raised major challenges in terms of the "learning environment and teaching methods".

Technology, particularly in the form of distributed or distance learning has dramatic implications for competition and for the role of the teacher. Abeles (1998) sees a split in the traditional roles of academics separating a few stellar researchers and lecturers, while the majority of faculty become mentors and guides to students. New partnerships are emerging to help cope with the enormous up-front investments needed and infrastructures such as those for quality assurance and intellectual property are being challenged. The changing technologies of research also have implications as yet not well-understood. On the one hand the cost of major facilities drives concentration as already noted (with an increasing tendency to locate these in universities rather than in dedicated service institutions in the public laboratory sector). On the other hand grid computing reduces the need for research to be co-located with such equipment. Likely changes in academic publishing may mean that the expense of subscriptions to gain access to large quantities of published information may vanish as public domain institutional repositories are demanded by research funders.

Collaboration with industry is taken as a central theme by the Forum which puts the problem as one of finding ways to link the world of “open science” with that of “open innovation” in a sustainable way that does not damage the long-term purpose of universities. The Swedish foresight study questions the ability of universities to engage with the Third Mission which is said to lack both legitimacy and funding (Heinegard 2005).

Transdisciplinarity in research is a recurring theme (Nowotny et al. 2001; Forum 2005 etc). Nowotny et al. explore the future in the context of the role of universities in knowledge production Mode 2. Among the trends they discern is one of deinstitutionalisation as the boundaries between universities and the world of commerce are eroded.

New Funding Patterns are also picked up as a driver of change. Some new funds derive from collaboration with industry and from commercialisation of research outputs, the extent of concentration of public research funding in the future and from which source it will come (national versus transnational for example) is another area of expected change. Tuition fees have been increasingly applied. The extent to which the individual rather than the state will pay for higher education remains a crucial policy question in many countries, including those which do not at present have them. Underpinning this driver is a pervading sense that the sector as a whole is under-funded.

These drivers in some cases are used to support scenarios for the future of universities. Table 9.1 summarises three examples, taken from the exercises discussed above. It should first be noted that they are produced by different means and in different contexts. Hence, the Strata-Etan group uses the exercise to work towards its third highly normative scenario which is in effect the conclusion of its report and is presented as a political priority. The CHEPS scenarios were produced by a specialist research centre in the HE policy and management domain on the occasion of its twentieth anniversary and emerged from a two round Delphi survey returned by 164 respondents across Europe. The OECD scenarios were built on an analysis of trends and aim to expose decision makers to strategic choices: they were in fact used as the basis for a discussion at a Ministerial conference (OECD 2006). In all cases the initial dimensions of design conceal a much greater complexity in what is envisaged.

Although they are constructed on different principles, there is an interesting convergence in the content of scenarios from all these sources and also some important distinctive elements. Thus on the first row of Table 9.1, a common element is the distinction between the public and private goods dimension of University activity, being maintained through a hybrid or quasi-market approach. On the second row the market dimension is emphasized, especially by the Strata-Etan group and OECD. This effectively presents a corporatisation of the university sector driven by full market forces. CHEPS do not go so far but nonetheless see the possibility of merger with private research institutions. In the second half of the Table 9.1 divergent scenarios are shown, all highlighting interesting aspects of possible futures.

Table 9.1 Comparison of scenarios

Strata-Etan	OECD University futures	CHEPS
Convergent scenarios		
Melting pot low socio-cultural & economic diversity combined with high social cohesion. Higher education system a hybrid similar to present producing undergraduate education and basic research as public goods and continuing professional education, applied R&D/ innovation in response to company and administrative demand. Scenario very difficult to manage but likely to occur if policy is laissez-faire in response to trends	New public management primarily publicly funded but greater use of NPM tools including market forces and financial incentives. Diversified funding including student fees. Students comparing teaching quality & employability. Marked division of labour but most still do some research. Research funding very competitive but mostly national expect for ERC. Accountability high	Centralia, the City of the Sun sees Europe in 2020 as the jolly old World with greying but rich leisured population. Universities are largely unchanged in function though many have combined as large merged or even national institutions. Blended mode learning combines campus with a network to make the most of ever smaller age cohorts. Fees are deregulated. Development & innovation that have not gone to Asia or Latin America have shifted to the South and East of Europe but remain linked to an elite D-University sector mainly in the North and West. Research is clearly separated between private & public goods
Market triumph neo-liberal economy and welfare crisis reduce social cohesion and diminish diversity, resulting in predominance of privatisation & marketisation with public goods and agents gradually losing importance. Results from system actively adapting to trends	Higher Education Inc. Global competition to provide research and education services. Research universities hardly teach and vocational ones hardly research. Demand driven except for most prestigious institutions where peer assessment remains. International division of labour in teaching and research with outsourcing of research to emerging countries. Highly concentrated research sector with fierce competition for stars	Octavia, the Spider-web City is a vision of multiple missions and visions while networks have become the main modes of coordination within and beyond universities. Successful universities capitalise on small units, thick information and multiple webs. Some have merged with private R&D facilities while others specialise around disciplinary or professional clusters. Research is organised in inter-faculty or inter-university units funded by national bodies, the ERC and international industry research consortia

(continued)

Table 9.1 (continued)

Strata-Etan	OECD University futures	CHEPS
Divergent scenarios		
Creative society results from proactive attitude within the HE system to adjust and support some trends but to resist others so as to become pillar of knowledge society. Focus is on public educational goods, with private goods supplied by companies. Research is focussed on basic research in interdisciplinary perspective, generic technologies and innovation in public utilities	1. Open networking HE very internationalised and involves intensive networking among institutions, scholars, students and with other actors such as industry. Model based more on collaboration than on competition. International modularisation. English as lingua franca and most standard courses online. Research collaboration with peer institutions 2. Serving local communities HE focussed on local missions and embedded in communities. Mainly publicly funded but a small elite retain international networking. Science mainly done in government sector and research seen as by-product in universities	Vitis Vinifera, the City of Traders and Micro-climates has Europe more focused on quality of life than innovation and the knowledge economy and the economy largely service based. HE is offered more flexibly by a wider set of institutions to a broader range of learners. The definition of universities is treated flexibly to encompass this diversity. ERC has partly displaced national funding of research, with funding highly selective and concentrated. Innovation is highly valued with much applied research now in universities

9.5 Tracing Policy Focus, Trends and Impacts

In assessing the impact of FTA in the HE sector we need first to identify the potential uses of FTA by different stakeholders in the system. For policymakers concerned with their innovation systems and/or the broader cultural economy there is a need to understand better the diversity and inner complexity of universities and HEIs as they strive to position themselves in the globalising knowledge society, characterised by increasing competition, new agents, student consumerism and demographic and technological impacts. Questions and issues include what types of support HEIs will expect from governments in terms of regulations, measures, new funding mechanisms? Even the legitimacy of government in setting the agenda could be called into question under some of the more laissez faire scenarios. On the other hand a stronger regulatory hand could be needed if new agents enter the system without the same levels of self-regulation observed by existing institutions. The long term spread of costs between taxpayer and consumer is also a core issue. In the specific case of innovation, the role of university research and training in supporting economic and social goals is a key issue. In part this is addressed by mainstream technology foresight activity but the institutional role of HEs may need to be addressed by more targeted exercises.

If we consider FTA as a policy instrument in itself, then the focus moves to its ability to act as an instrument of change. There is some evidence that governments, and some other stakeholders such as industry, are not satisfied with the current performance of their HE sectors. International league tables have served to highlight the deficiencies of some systems. FTA offers the possibility to explore alternative scenarios and to identify visions of where the system should be going as a first step to embarking upon a path of change and development.

The more advanced HEIs which are emerging as world class centres in specialised areas are oriented more to the needs and competitive pressures of the global economy. Their operation transcends national boundaries and is the concern of multilevel governance from an equity and human and social development perspective. Similarly the delivery of distance courses to large numbers of students worldwide raises the need for multilevel governance scrutiny, regulation and support measures. What form could this take? As outlined above, international organisations like OECD, World Bank, UNU and UNESCO are positioning themselves increasingly in partnerships, to take up these challenges.

Depending upon the scope of the question all of the above-mentioned issues could be raised at local/regional, national or transnational level.

If we move to the needs of HEIs themselves, they could be said to be seeking space to self-organise and identify the appropriate positioning based on their core competencies and current strengths whilst also responding to emerging local needs and priorities. If we consider the examples discussed in this paper the impression is that FTA here is strongly embedded in the strategy-making process of the organisation. It is no coincidence that the examples come from countries where HEs have enough freedom of manoeuvre to develop effective strategies. This situation is both a strength and a weakness. On the one hand it ensures an impact, but on the other it may also create constraints. One such constraint is path dependency – a strategy must always start with the current position but may also import the assumptions which led to that position. The other constraint is the risk of being inward looking and unwilling to engage sufficiently with external stakeholders. The institutions discussed, Melbourne, Manchester and DCU all have exercises led from the top, which increases the chance of impact but then creates the challenge of getting buy-in to the process from staff and students. All three are making substantial efforts to do this.

At a sectoral level FTA is problematic both because of the existing diversity of institutions and even more so because most of the futures we have considered imply an increasing diversity in the future, and an emergence of elites, which will propel a significant number into a level which does not flatter their self-image or ambitions.

At the national and international level we encounter the broader problems of ensuring impact of FTA. Earlier work has suggested that a key factor here is the extent to which the FTA activity is embedded in the implementation environment (Georghiou and Keenan 2006). In other words, when FTA activity is planned are those who are in a position to use the results engaged with the exercise? What other influences are likely to affect their behaviour and how important are these relative to FTA findings?

These challenges emerge quite clearly when we consider the international FTA activities. They have an increasing legitimacy as policy for HE becomes internationalised through public initiatives such as the Bologna process and the European Research Council and putative European Institute for Technology (European Commission 2006). Internationalisation is also driven by the competitive and collaborative trends we have discussed at length. However, at least the European studies only seek to inform debate – they have no committed audience. Of the international organisations, OECD has a more formalised route to policymakers though less so to the sector itself. The latter, together with the World Bank and the UN agencies, can collectively through FTA mobilise major changes in HE in developing countries with obvious impacts on the developed world. Nonetheless, transnational FTA activity is likely to have an important role in policy transfer between countries.

9.6 Conclusion

This paper has mapped and reflected on the range and diversity of FTA activity undertaken in recent years from a global perspective and largely sponsored by international organisations. The insights and results generated by these studies highlight a growing concern with emerging trends in higher education and alert the reader to the potentially negative impacts of these trends on those less advantaged active in the HE sector as a supplier or consumer of the services. International FTA poses a clear challenge in reviewing/ assessing the effectiveness of such studies on the basis of the extent to which this work helps in shaping and producing more desirable or preferred futures for the sector. To date FTA has undoubtedly helped to create an enhanced awareness and to spark a debate on the differing perspectives on and alternative scenarios for the future of the HE sector. As is often the case with FTA, effects may be delayed as reports enter a "reservoir" of knowledge which is drawn upon in response to an emerging strategic or policy challenge. As the sector faces growing uncertainty on the one hand and closer policy attention as an instrument of the knowledge economy on the other, it is likely that the demand for foresight visions from FTA will grow. With so many stakeholders it is less clear at which level an authoritative voice will emerge.

References

Abeles T (1998) The Academy in a Wired World, *Futures* Vol. 30, No. 7, pp. 603–313

AVCC (2002) Survey of Applicants for Undergraduate Higher Education Courses, AVCC Annual Survey, Canberra, available at: http://www.avcc.edu.au/content.asp?page=/publications/stats/unmet/index.htm

Cassingena HJ and Georghiou L, (2005) Foresight in Innovation Policy: Shared Visions for a Science Park and Business – University Links in a City Region, *Technology Analysis & Strategic Management* Vol. 12, No. 2, pp. 147–160

Coombs R and Georghiou L (2002) A New Industrial Ecology, *Science* Vol. 296, p. 471

Cuhls K and Georghiou L (2004) Evaluating a Participative Foresight Process: Futur – The German Research Dialogue, *Research Evaluation* Vol. 13, No. 3, pp. 143–153(11)

Dearing R (1997) Report of the National Committee of Inquiry into Higher Education

European Commission (2006) Implementing the renewed partnership for growth and jobs. Developing a knowledge flagship: The European Institute of Technology, Communication to the European Council February 22 COM(2006) 77 Final

Etzkowitz, H., Webster, A., Gebhardt, C. and Terra, BRC. (2000) The Future of the University and the University of the Future: Evolution of Ivory Tower to Entrepreneurial Paradigm, *Research Policy*, 29, pp. 313–330

Forum for University-based Research (2005) *European Universities: Enhancing Europe's Research Base*, European Communities

Georghiou L (2001) Third Generation Foresight: Integrating the Socio-economic Dimension, *NISTEP Study Material No. 77*, pp. 235–244

Georghiou L and Keenan M (2006) Evaluation of National Foresight Activities: Assessing Rationale, Process and Impact, *Technological Forecasting and Social Change* Vol. 73, No. 7, pp. 761–777

Glenn J and Gordon TJ (2004) Future S&T Policy Management Issues – 2025 Global Scenarios, Technological Forecasting & Social Change Vol. 71, pp. 913–940

Green M, Eckel P and Barblan A (2002) *The Brave New (and Smaller) World of Higher Education: A Transatlantic View*, European University Association and American Council on Education Center for Institutional and International Initiatives

Havas A (2008) Devising Futures for Universities in a Multi-level Structure: A methodological experiment, *Technological Forecasting and Social Change* Vol. 75, pp. 558–582

Heinegard C (2005) The University of the Future, IVA (Swedish Royal Academy of Engineering)

Juma C and Yee-Cheeong L (coordinators) (2005) *Innovation: Applying Knowledge in Development*, UN Millennium Project Task Force on Science, Technology and Innovation, Earthscan, London and Sterling, Va

Kaiser, F., Boezerooij, P., Huisman, J., Jongbloed, B., Kaulisch, M., Luijten-Lub, A., Nokkala, T., Salerno, C., Theisens, H. and Vossensteyn, H. (2004) Higher Education Policy Issues and Trends. An Update on Higher Education Policy Issues in 11 Western Countries, Higher Education Monitor Update Report, 2004, Twente, CHEPS

Kirp D. (2003) Shakespeare, Einstein and the Bottom Line – The Marketing of Higher Education, Harvard, Cambridge

Klusacek K (2004) Technology Foresight in the Czech Republic, *International Journal of Foresight and Innovation Policy* Vol. 1, Nos 1–2, 2004

Lambert R and Butler N (2006) The Future Of European Universities: Renaissance or decay? Centre for European Reform, London, June 2006

Nedeva M, Georghiou L, Loveridge D and Cameron H (1996) The Use of Co-Nomination to Identify Expert Participants for Technology Foresight, *R&D Management* Vol. 26 No. 2 pp. 155–168

Nowotny H, Scott P and Gibbons M (2001) Rethinking Science – Knowledge and the Public in an Age of Uncertainty, Polity Press, London, p. 278

OECD/CERI (2005) University Futures Project: Six Scenarios for Universities – a discussion paper

OECD (2006) Meeting of OECD Education Ministers, Higher Education: Quality, Equit and Efficiency, Athens 27–28 June

Skolnik M.L. (1998) Higher Education in the 21st Centrury – perspectives on an emerging body of literature, *Futures* Vol. 30, No. 7 pp. 635–650

Slaughter R (1998) Universities as Institutions of Foresight, *Journal of Futures Studies* Vol. 1, No. 3, pp. 51–57

Strata-Etan Expert Group on Foresight for the Development of Higher Education/Research Relations (2002) *Higher Education and Research for the ERA: Current Trends and Challenges for the Near Future* – Final Report of the Strata-Etan Expert Group

Thorne M (ed) (1999) *Foresight – Universities in the Future*, UK Department of Trade and Industry, DTI/Pub 4263/5k/7/98/NP URN 99/982

University of Manchester (2006) Towards Manchester 2015 – the Strategic Plan of the University of Manchester available at: http://www.manchester.ac.uk/medialibrary/2015/2015strategy.pdfuture

Vincent-Lancrin S, (2004) Building Future Scenarios for Universities and Higher Education: An International Approach, *Policy Futures in Education* Vol. 2, pp. 245–263

Vincent-Lancrin S, (2006) What is Changing in Academic Research? Trends and Futures Scenarios, *European Journal of Education* Vol. 41, No. 2

Chapter 10
Applying FTA Methods in Latin American Countries

M. Albornoz

10.1 Introduction

Future-Oriented Technology Analysis (FTA) methods have been used in highly developed countries for the last decades to improve competitive strategies, evaluate social responses to technological development paths and foresee critical situations in sensitive areas, such as energy, the environment, natural resources and demographic trends, as well as to explore the implications of advanced technologies in areas like ICT technologies and basic research on molecular biology.

FTA methods have been increasingly considered as essential inputs for decision-making at different levels, whether by national and regional policy makers or by businesses and other actors. These methodologies now play an important role in decision-making in the fields of R&D, risk evaluation, environmental studies, S&T policies and strategies design, support for innovation, and evaluation of the social impact of innovations.

As the nature of these decisions become more complex, so FTA methods are becoming increasingly sophisticated and diverse. Methods for analyzing future technologies now include different families, such as statistical and trend analysis, model and simulation, expert opinion (including the Delphi method), mathematical simulation models and scenarios, among others.[1] The common feature of all these methods is that they are tools for ordering information, discovering behaviour patterns, and understanding relationships between different phenomena, thus making the social impact of developments in the field of science and technology more foreseeable.

The value of FTA methods for formulating accurate predictions of future events is now widely accepted by decision makers in both the public and the private sectors. This point is worth stressing because the effectiveness of FTA methods in decision making processes not only depends on the quality of the studies produced,

[1] Michael Rader and Alan Porter propose a typology of 48 methods arranged into 13 "families", based on a previous typology consisting of 51 methods arranged into 9 "families" identified by the Technology Future Analysis Methods Working Group in 2004.

C. Cagnin et al. (eds.) *Future-Oriented Technology Analysis,*
doi:10.1007/978-3-540-68811-2,

but also to a large extent on demand for these studies from decision makers, both in the public and private sector. Here, as in other fields of social activity, demand plays a decisive role.

FTA instruments could also be very useful for less developed countries (LDCs). As they face major challenges from global competition and current paths of technological development, FTA could play a key role in both areas in designing strategies for social and economic development aimed at taking advantage of opportunities and reducing risks. Nevertheless, experience in this field is limited and there are still difficulties in applying FTA instruments in the economic, social and political context of LDCs.

However, no LDC can show sufficient experience in this field as to take advantage of the contributions of FTA methods to make more rational and accurate decisions in areas related to social, economic and technological development. Of all the Latin American countries, Brazil is the country with the greatest experience, yet it only began to develop these activities less than a decade ago. Most Latin American countries, like most other LDCs, have only recently started to apply FTA methods. However, the fact remains that in the 1970s Latin American countries carried out some highly ambitious forecasting exercises, such as the Bariloche Foundation's Latin American World Model, created in response to the "Limits to Growth" report. However, there has been little follow-up since then on these pioneering exercises.

This chapter focuses on experiences with FTA methods in Latin America. It is possible to identify two different stages in the use of FTA in the region. The first, in the 1970s, had a holistic approach and was related to a Latin American view of development. The second, from the present decade, has a closer focus on specific problems and opportunities, within the framework of increasing globalization and the advent of the knowledge society. It remains to be seen whether the findings presented here are applicable to other LDCs outside the region.

10.2 First Latin American Experiences in Future-Oriented Studies

Argentina, Brazil and some other Latin American countries were forerunners in the use of future-oriented methodologies for improving policies in the area of S&T. The key word was "prospective" as a synonym of foresight. The two pioneering prospective studies were the Latin American World Model and the Technological Outlook for Latin America project (known in Spanish as PTAL).

The Latin American World Model was designed by the Bariloche Foundation of Argentina, under the direction of Amílcar Herrera in the early 1970s (see Herrera et al. 1976), with the purpose of refuting the Club of Rome's "Limits to Growth" report and creating a model based on alternative assumptions (Marí and Callejo 2000). The "Limits to Growth" model was an analysis that set out to show the limits imposed on growth by the physical environment. Based on various assumptions about the behaviour of humanity and the availability of natural resources, its central hypothesis was that exponential population growth and consumption would necessarily lead to a "catastrophe" midway through the twenty-first century. Unless these trends were corrected, the depletion of non-renewable natural resources and environmental pollution would eventually lead to the collapse of the ecosystem. The key to averting the catastrophe envisaged in the Club of Rome's report was to control population growth, reduce pollution and use resources in a rational way.

The Latin American World Model started from a different assumption: the most important problems confronting the modern world are not physical but social and political, and arise from the unequal distribution of power in the world at both national and international levels. Consequently, it did not try to predict trends by examining present realities, but to propose a final goal, a wished-for scenario or the image of an ideal society. At the preliminary stage, discussion was centred on the theoretical assumptions in the "Limits to Growth" report about the availability and use of natural resources and increasing pollution. The concept of "reserves" was also discussed – a concept that did not reflect the earth's riches, but only known resources at that time. The authors of the model indicated that although there were no scientific reasons to anticipate an ecological catastrophe or an acute shortage of resources as "Limits to Growth" has done, this did not mean that such danger could not exist if the social model changed. On the contrary, they argued that the social model they proposed guaranteed that there would be no danger of a catastrophe. They also argued that it was necessary to reduce the rate of technological development since advances in technology had already outstripped existing consumer needs.

An attempt was made to show with a mathematical simulation model that the expenditure of natural resources was not the real problem and that, as things stood, the different countries or regions of the world, especially the poorest, could achieve the proposed goal in a reasonable period of time. Various applications of the mathematical model demonstrated that by applying the proposed policies, mankind could reach acceptable levels of well-being in just over a generation, with practically no physical limitations.

With regard to the assumptions mentioned earlier and the idea that there are no absolute physical limits to human development, the model tried "to demonstrate" the viability of the proposed society starting from the existing natural resources, but assuming that the necessary social and political changes required by the model would actually happen. The model also showed that population growth would descend as general living conditions improved, especially those related to basic necessities. The Latin American World Model did not try to diminish the problem of the exhaustion of resources. Simply, it wanted to focus the model on social and political structures. The development of the model was a valuable experience because, in addition to its intrinsic value, a new school of technicians was created that spread across Latin America. In particular, their model served as an inspiration for a long term economic simulation model adopted by the United Nations. This model was used for the development of long term planning models and for the formation of technicians (Marí and Callejo 2000).

A spin-off from the Latin American World Model was the Technological Outlook for Latin America project (PTAL) that started up in 1983. Again,

Amílcar Herrera led the project but this time it was located at the Centre for Scientific and Technological Policy of the Institute of Geosciences at the University of Campinas (Brazil) and a network of Latin American centres took part (Marí and Callejo 2000). The project came at a moment when the impact of the new technologies spreading across the globe was threatening to change existing patterns of production and theories of technical change. The project analyzed these trends and existing future-oriented studies. Like the Bariloche Model before it and unlike the trend extrapolation studies that dominated in the developed world, the project took an ideological and normative approach. As in the Bariloche Model, an ideal society was defined: egalitarian, participative, independent (non-autarkic), with free time for creative activities, sober, intrinsically compatible with the physical environment.

The Outlook for Latin America (PTAL) project was methodologically different, from the Latin American World Model. While the latter had been based on simulations, PTAL worked with scenarios (Herrera et al. 1994). In order to construct the scenarios, key variables, both internal and external, were defined. Attempts were made to integrate some external variables into bipolar constructs, such as the tension between dependency and autonomy in Latin American countries, or alternative scenarios of world-wide economic crisis in contrast with a new phase of global economic growth. The internal variables included the predominant type of social agreement, the level of Latin American cooperation that could be achieved and the style of development. Also transformation factors were considered, such as currently existing "heavy" trends and the germs of the future (fundamentally the new technologies and the emerging social movements). The project assigned great importance to social movements as a fundamental element of change.

The project devoted much of its efforts to the analysis of impacts and possibilities arising from developments in new technologies, mainly computer science, biotechnology and new materials. To this end, guidelines for scientific and technological policies were suggested in each of these areas. Unlike what had happened with the Latin American World-wide Model, the PTAL not only tried to design an ideal scenario, but also a strategy to achieve it.

The first prospective studies were intended to understand and predict the structure of the world, as well as to support an alternative vision of the international distribution of power and wealth. For this to materialize, it would have required the capacity to introduce major changes in the economic and political international order. This utopian condition affected the predictive power of the models by depriving them of an objective point of view. As the development model followed by Latin American countries during the 1970s began to fail and the world economy entered a period of transformation, the wishful thinking that had inspired these models became increasingly obvious.

These were not, however, the only experiences, although they were the most important in terms of ambition and the scope of their undertaking. There were also experiences in Colombia from the late 1960s. In 1969, Colciencias initiated the Operation Development and Colombia Group Year 2000 Projects, which aimed to promote long-term studies. Chile also saw early work in this area, led by academics from different national universities. In general, these experiences had high academic relevance and represented notable creativity and reflection. However, they did not meet demand from specific decision-makers at policy or business level.

10.3 Recent Experiences

Since the start of the present decade there has been a common interest in future studies, although on this occasion the theoretical and methodological assumptions are different to those previously seen. It is important to note that there has been a hiatus, not only temporal but methodological and ideological, between one period and the other. These new experiences are connected to trends seen in developed countries and include academic consultancy and support from leading groups in this area. However, in general terms the phenomenon is incipient and has yet to establish itself in the region. It must be taken into account, as indicated above, that the strongest country in this field, Brazil, has only been involved for a little over five years. Indeed, the Centre for Strategic Studies and Administration (CGEE in Portuguese) was created in 2001 to support decision-making processes, and the formulation and implementation of public science, technology and innovation policies. CGEE activities involve a group of actors, including management and technical staff, specialists, policy-makers and decision-makers. The centre interacts at the principal stages of construction and implementation of public science technology and innovation policies, in order to add value to its foresight studies. The CGEE's positioning in relation to decision-making has made it possible to promote synergy and provide support for the formulation of public science technology and innovation policies.

Other experiences in Brazil date from the same period. Such is the case of Prospectar, a study implemented by the Ministry of Science and Technology (MCT in Portuguese) as a nationwide exercise, carried out with the Delphi methodology. The first phase of Prospectar took place between 2001 and 2002 and examined eight areas: aeronautics, farming (including forestry and fishing), energy (including biomass), space, materials, water resources (including river transport), health and telecommunications/information technology (including electronics). The Brazilian Technology Foresight Program, created by Brazil's Ministry of Development, Industry and Foreign Trade, has focused on future-oriented analysis of chains of production in the plastics industry, civil construction, textiles and clothing, wood and furniture, among others.

In 2002 Colombia created the National Program for Technological and Industrial Foresight (PNP). The purpose of the program was to increase national capabilities and encourage dialogue between the government, industrialists, scientists and engineers and other social actors in order to construct a shared vision of the future and design long-term policies and strategies for scientific, technological and industrial development. The main organizations taking part in the program are COLCIENCIAS, the Andean Development Corporation and the Ministry of Industry, Commerce and Tourism. There is a permanent manager and a national committee made up of representatives from different institutions (Medina Vásquez and Rincón Bergman 2006). Colombia has also relatively recently implemented Science and Technology Foresight Agendas, promoted by different departments and regions of the country. Futures studies have also been carried out in production chains, such as plastics, dairy products, leather and packaging. In addition, twelve different types of intervention in foresight have been developed, in areas such as biotechnology, energy, health, tourism, centres of excellence, international cooperation, knowledge economy and society, among others.[2]

Chile has a Technological Foresight Program (PPT) belonging to the "Program for Development and Technological Innovation" run by the Ministry of the Economy together with the Corporation for the Promotion of Production (CORFO), the National Commission for Scientific Research and Technology (CONICYT), the Foundation for Agricultural Innovation (FIA), the National Standards Institute (INN) and the Corporation for Technological Investigation (INTEC-Chile). The well-known study "Imagining Chile's Economic Future" examined four fields: wines, aquaculture, biotechnology, and education. In Argentina, The National Institute for Scientific and Technological information (CAICYT), belonging to the National Research Council (CONICET) organized in 2004 an area for strategic intelligence studies and services.

Peru has an Office for Innovation and Technological Foresight, part of the National Science and Technology Council (CONCYTEC). Active since 2001, the Office's roles include evaluating multi-institutional development and technological innovation projects and consulting with shareholders in creating technological innovation and foresight plans and programs. In addition, in 2001 the Ministry of Science, Technology and Environment of Cuba created the Cuban Science and Technology Observatory (OCCyT in Spanish), which aims to identify scientific and technological trends and draw attention to results achieved internationally in this area. Likewise, the OCCyT makes evaluations of the Cuban science and technology system, analysing and evaluating strategic development issues in this area in relation to national economic, social and environmental priorities.

These examples do not constitute an exhaustive list of what has occurred in all the countries of Latin America in this area, but they paint a fairly accurate scenario of

---

[2] Work by Javier Medina Vásquez reports on Colombia's experience in applying FTA in the Andean zone. It shows that some working groups in universities and technology centres have developed strengths in several regions of the country. The main national S&T institution, Colciencias, executed a foresight program between 1986 and 1990, and during the following years different activities were carried out. In 2001 and 2002 an incubation period took place, with activities sponsored by UNIDO. Then, by the end of 2002 Colciencias, the Ministry of Commerce, Industry and Tourism, and the Andean Corporation for Promotion created the Colombian Program for Technological and Industrial Foresight. In 2005 and 2006 efforts were made to orientate foresight skills to identify and support strategic sectors. The Program has also established successful cooperation with the Technological Foresight Project of the Convenio Andrés Bello, an international organization in which twelve countries participate in science, technology, education and culture activities (for more information see Medina Vásquez 2006).

growing interest in an area that remains undeveloped on the whole. To complete the scenario is it interesting to examine the types of methodology most frequently used.

10.4 Applying Different Methodologies

The different FTA methodologies include different methods and disciplinary approaches, each of which has a set purpose or can be used more suitably in some situations than in others. This is not the only way in which they differ, as they also differ in the type of resources they mobilize and in the technical conditioning of their application. Michael Rader and Alan Porter (see Chap. 3) have grouped the different methodologies into 13 "families": (a) creativity approaches, (b) monitoring & intelligence, (c) descriptive, (d) matrices, (e) statistical analyses, (f) trend analyses, (g) expert opinion, (h) modelling & simulation, (i) logical/causal analyses, (j) roadmapping, (k) scenarios, (l) valuing/decision aiding/economic analyses, and (m) combinations of different methods from the different families. Some of these methodologies, which can be described as "soft", are based on the use of experts' know-how. Others, which can be considered "hard", are based on simulations and models or on the use of databases, and therefore require technical resources and specific skills not always available in developing countries. It is no surprise, then, that the most used methods in most of Latin America are Delphi surveys and scenarios. Experience of other methodologies is more limited, with the exception of mathematic models which developed considerably in the first period.

Monitoring & Intelligence methods are not widely applied to science and technology policy making in Latin American countries, due to a lack of institutional technical capabilities. Technological Intelligence (TI) consisting of methods gathering reliable information and converting it into an intelligent product for decision-making (Escorsa et al. 2006) requires the capacity of accessing and exploiting international bibliographic and patents data bases. The tools of TI are similar to those used in bibliometry: text mining or data mining and the creation of conceptual "maps" of patents or lines of research and development by means of suitable software. This brings up the question again of the need for a minimum institutional infrastructure, such as competitive strategy centres. These exist in very few countries in Latin America. Descriptive methods, comprising bibliometrics and other methods, have a recent but growing presence in Latin American countries. The National Institute for Scientific and Technological information (CAICYT) in Argentina, for example, has developed a bibliometrics and patent studies and services area, and is developing its own search engines for bibliometric data mining. Other institutions, such as BIREME in Brazil, which promotes a Latin American platform of full-text, open-access scientific journals, also has a highly relevant experience in developing and using bibliometric tools.

Expert Opinion methods, which include surveys, Delphi, focus groups and participatory approaches, have been the most frequently used in Latin American countries, although their application has not necessarily been extensive. Some initiatives have been mainly limited to a small number of Delphi studies oriented to improve centralized planning, mostly at a national level. Indeed, the majority of recent future-oriented studies in Latin America in the field of science and technology have been based either on the Delphi method and or in some cases, on scenario writing. One of the reasons why Delphi has become the method of choice for FTA in Latin America is its flexibility, since it only requires experience in carrying out surveys. Unfortunately, it is not always possible to balance a Delphi study with more "objective" methods as this type of information is frequently not available.

Nevertheless, there have been several other national initiatives. In Brazil some Delphi surveys have been aimed at enhancing competitiveness and technology training in industry, improving the quality of life of the population and promoting strategic R&D lines. Delphi forecasting has also been used in human resources training and to develop Brazil's scientific capacities and technological infrastructure. In Colombia, four Delphi surveys with a number of experts have been implemented (Medina Vásquez 2006).

Chile organized a Delphi study of the country's positioning strategy, priorities for product development, training and education needs, technology development needs and other strategic areas. In 2001, Uruguay implemented the "Uruguay 2015 Technology Foresight Program" within the framework of a wider national foresight Initiative. This program focused on areas such as logistics and transport and used Delphi and scenario methods. In 2002 a study was made of biotechnology in the agro-food system, using panels of experts. Argentina also carried out a Delphi study as the basis for its strategic plan in STI 2005–2015.

In 1999 the OAS financed a project bringing together experts from six different countries: Argentina, Bolivia, Brazil, Chile, Paraguay and Uruguay. The project, Opportunities and Strategies for the MERCOSUR, involved the use of the Delphi method on a regional scale. The study included defining strategic targets, identifying opportunities, analyzing trends and constructing future-oriented hypotheses regarding the extended MERCOSUR (Chile and Bolivia were then in the process of joining the four existing member countries). The timescale of the study was limited to five years into the future because of the volatile nature of economic, social and political situations in the region.

The project was developed by an international team of specialists in scientific and technological policy from each of the member countries, plus Chile and Bolivia. The first round included 120 significant actors involved in regional decision making processes (politicians responsible for coordinating MERCOSUR activities, opinion leaders, as well as spokespeople in important positions for the process of regional integration). In the second round, significant actors in the field of science, technology and innovation were interviewed.

The first Delphi exercise aimed to identify strategic elements for the development of the MERCOSUR, taking the views of important social actors involved in the process of regional integration as its starting point. Participants answered questionnaires designed to elicit their views about the future of regional integration. The data obtained made it possible to define strategic areas and opportunities for

MERCOSUR development, as well as giving an overview of the current level of integration of scientific activities within the process of regional development.

Once high-priority areas and opportunities had been identified in the first round of consultations, a second round was carried out to determine the extent to which the national and regional scientific capabilities needed to achieve the previously identified strategic targets were actually available. In particular, questions focused on the availability of these capabilities, the need to develop new capabilities, the extent to which different national science and technology systems in the region complemented one another, the potential for intra-regional scientific cooperation initiatives, and the number and effectiveness of existing policy instruments.

Not only scientists and technologists participated in this second round of consultations, but also other actors playing a significant role in the definition of science, technology and innovation policies: innovative entrepreneurs, science and technology policy analysts, and technology consultants. Around 60 specialists from the different member countries were consulted. Additional interviews were held with specialists in the previously determined strategic areas. Members of the scientific and technological communities and the economic-industrial sectors were asked to define how the different national systems of science, technology and innovation might contribute to the MERCOSUR's strategic goals.

The recommendations of the study were adopted by the MERCOSUR's Special Meeting on Science and Technology (known in Spanish as RECyT), the body responsible for promoting regional integration in science and technology and defining the guidelines for integration in this area. A regional "Framework Program" for science, technology and innovation is currently being drafted based on recommendations from this study.

Modeling & Simulation methods include innovation system descriptions and complex adaptive system modeling, among others. The former method is growing rapidly in several countries, due to the fact that the innovation systems approach is currently the new framework for science and technology policies. Moreover, the design of mathematical models to represent complex systems is a technique that has been used in Latin American countries to represent complex global, regional and local systems and to predict how they might evolve. The most outstanding example of this has been the above-mentioned Latin American World Model. It was not the only one, however. In 1995, a regional project known as "The ecological future of a continent. A prospective vision of Latin America" used simulation models to study land use within alternative socioeconomic scenarios.

The development of mathematical models for studying social processes in LDCs can be traced back to Oscar Varsavsky (Varsavsky and Calcagno 1971), whose approach was relatively widespread in Latin America some years ago, particularly in Buenos Aires, Caracas and Santiago de Chile. The method, known as "numerical experimentation" applied mathematical models to the social sciences. However, its usefulness was limited by the fact that the models can only detect clearly understandable and quantifiable elements and relationships, which can prove to be problematic in Latin America. In the field of science and technology, Latin American countries have many difficulties in developing long series of indicators. A basic shortage of information affects the validity and reliability of the models and accentuates the subjective biases of the people who design them. In addition, a minimum number of experts in mathematical modelling of complex systems is required, as well as a suitable institutional context. Few centres in the region have the necessary capability at present to develop models of this type and there is very little consensus on the need for them.

Scenarios, as a method for extrapolating existing information into alternative versions of the future and for understanding the logic of events, are frequently used in Latin American countries, particularly to analyze regional development and ecological problems. It is a relatively easy technique to apply since all it requires is the presence of experts in the field to be studied and the ability to organize the different dimensions in a logical manner. The strengths and weaknesses of this method are closely bound up with the logical structure of the scenarios: if the logical choices are well formulated, then it is a powerful tool; otherwise, it simply goes round in circles. Scenario writing is generally used to complement Delphi surveys.

In the 1980s, the FAST Program, supported by the European Union, commissioned and financed a study of Scenarios in Latin America which was carried out in 1989 and coordinated by the Centre for Advanced Studies of the University of Buenos Aires. Academics from Argentina, Brazil, Chile and Uruguay participated in the program. From the methodological point of view, alternative scenarios were constructed on the axes of "growth" and "equity". The project established a typology of scenarios regarding dimensions of political structure, economic policy model, directions in higher education and science and technology policy. The scenes were evaluated using a face-to-face Delphi-type survey with leaders from several Latin American countries. In Colombia, COLCIENCIAS gave fresh impulse to future-oriented studies by financing the foresight agendas of regions like San Andrés, the Guajira, Amazonas, Casanare and Antioquia, among others. The agendas set out to analyze scenarios for regional technological scientific development with time horizons of ten and twenty years.

The list of cases may give the impression than there has been a broad accumulated experience. However, most of these are isolated and many have been discontinued or their recommendations ignored. The community of experts in FTA methods is small and has insufficient consolidated, permanent, incremental centres with an installed analytical capacity where learning can thrive. The lack of demand from governments or businesses conspires against the growth of most existing centres. In Argentina, this situation is slowly beginning to change as science and technology bodies and some businesses start to generate specific demand in areas such as nanotechnology, biotechnology and biofuels.

10.5 Difficulties for FTA in Less Developed Countries

Which could be the explanation for the phenomenon of LA countries having been forerunners in the past and being "followers" or "incipient" users of FTA methods in the present time? How can the gap be explained? Some explanations are related to the context: the world has changed. Latin American initiatives were infuse with notions of endogenous development, which called into question the emerging political and economic order of the post-war period. Globalization and the increased path of technological change were some of the elements that caused a deep crisis of that development model. So far, Latin American countries cannot find successful strategies to face the new challenges. As somebody in Latin America once said: "in the good old times, the future was better". At least, it seemed to be more predictable.

Plausible explanations for the paradox are mostly related to the structural conditions of LDCs. As Bell (1994) pointed out, the basic conditions for "prognosis", as he calls future oriented studies, are an abundance of available information, and the existence of stakeholders with foreseeable behaviours. These factors, in turn, depend on the existence of appropriate social structures and adequate management resources, which are often lacking in LDCs.

Indeed, foresight studies do require stable situations and recurring phenomena or trends whose general direction, at least, can be described in terms of statistical time series or persistent historical tendencies (Bell 1994). In many LDCs these statistical temporal series are not available. In some countries, specific FTA methods cannot be applied because of poor surveys and lack of systematic information in basic areas, as well as difficulties in accessing international databases. The cost of accessing bibliometric databases, for instance, is an almost insurmountable difficulty for S&T institutions in LDCs. Many countries have severe difficulties in collecting and processing basic indicators of science, technology and innovation in a reliable and continuous way owing to a lack of expertise in the institutions responsible for producing that kind of information.

Foreseeable stakeholder behaviour is the other necessary condition for sustaining future studies. Foresight in social, economic and technological processes is only possible when it can be assumed that the people who influence events behave rationally, recognize the limitations on their actions and the costs attached to different courses of actions, are prepared to play by the rules as these are generally understood and are prepared to be consistent. The fact that the behaviour of the stakeholders is difficult to predict is partly due to lack of social, economic and political stability in LDCs, and partly to the fact that multinational corporations with factories in LDCs, are based abroad and their decisions are influenced by global, rather than local, concerns.

However, the biggest difficulty is structural in nature and derives from the fact that LDCs are not leaders but followers of technologies developed in other contexts. Thus, FTA methodologies developed in countries working on the S&T frontier are not always appropriate for LDCs. This explains some of the criticisms that have been made about the validity of FTA methods and suggests there is a need to create new methods or adapt existing ones to the needs and operative possibilities of LDCs. Some inspired thinking is needed in order to make FTA tools available to LDCs.

There is a fundamental difference between the forecasting done in developed and developing countries: in industrialized countries, forecasting is linked to the pushing back of existing frontiers, the potential offered by knowledge and the future demands of society. In LDCs forecasting takes on a different meaning. Faced with an international scenario of fierce scientific and technological development and increasingly limited resources, it is necessary to carefully evaluate the opportunities and the threats.

Another characteristic of developing countries is that they do not make the decisions; they receive them passively. Consequently, except in certain niche markets, forecasting is limited to monitoring those areas that show the greatest potential. Nevertheless, this subject is more complex than it might seem at first sight since it is not just a problem of information. Underdevelopment implies structural conditions that are very vulnerable to global events, which are difficult to anticipate and even more difficult to correct.

Universities, for example, are preparing many of their best graduates to cover the demand for professionals in industrialized countries. How they can make long term decisions in a context like that? In short, uncertainty is greater in LDCs, there is less information available and fewer variables can be anticipated and controlled.

Nevertheless, these structural limitations do not imply determinism, and there are always opportunities that LDCs can explore and strategies they can adopt to integrate more successfully into the global economy. This gives FTA an additional task, namely, to identify such opportunities. The ability to do this would offer LDCs a powerful instrument to plan their development.

10.6 The Challenges

The main challenge is to promote changes in decision-making processes in order for decisions to be based on high quality information about their future consequences and impacts. Such a task is naturally beyond the scope of experts in FTA since it would require changes in behaviour from other high-ranking actors in public and private institutions. Nevertheless, even within the field of FTA studies it is possible to demonstrate the need for this sort of information to define the available options more clearly and understand the consequences of present decisions.

Additionally, following the previous statements, some challenges are related to building basic capabilities on at least three levels:

1. Information systems and access to databases
2. Links among heterogeneous actors
3. Human and institutional resources

1. Information systems

Information systems are a major area for improvement in Latin America and in most LDCs. Basic indicators in the fields of science, technology, innovation and even higher education lack the necessary continuity and standardization. It is also necessary to facilitate access to bibliometry and patents databases for smaller countries with fewer available resources.

2. Linking heterogeneous actors

The second area for improvement – probably the most essential one for LDCs – is networking. Systemic links should be developed among heterogeneous actors, such as academic groups and stakeholders, both at government and businesses levels to streamline FTA studies to decision-making.

3. Human and institutional resources

A certain number of trained people are needed so that FTA tools could be efficiently used. This means providing training and support for groups carrying out foresight studies.

The need to develop capacities implies the importance of promoting networking. For instance, training activities could be enhanced by networking between experienced and less-experienced groups, which would help to spread know-how and allow groups to learn from each other's experiences. Very few future oriented studies are carried out at postgraduate level in Latin American universities. To make up for this scarcity, regional postgraduate programs are needed to pool the existing capabilities of different countries and thus contribute to the training of professionals in this area. A network of postgraduate students might compensate for the current shortage of professors with necessary skills.

Indeed, Latin American countries have in recent years developed interesting networking experiences in areas of science, technology and innovation policy. The Foresight Network of the Ibero-American Program of Science and Technology for Development – CYTED, is a good example of this. The network, which began operating in 2003 and is currently active, was created to promote regional capabilities in technology forecasting and to serve as an instrument for formulating government, institutional and company policies and strategies. It is made up of experts from Argentina, Bolivia, Brazil, Colombia, Cuba, Spain, Uruguay and Venezuela. One of the objectives of the network is to promote forecasting in areas of interest to the CYTED Program and members countries.

Another experience comes from the Latin America and Caribbean Technology Foresight Initiative supported by the United Nations Industrial Development Organization (UNIDO). It was created as a network of Latin American and Caribbean countries with the purpose to exchange knowledge on technology forecasting between these and other countries. The program, which was only partly implemented, set out to construct a database of experiences and methodology in this field and to provide opportunities for distance training. Although Brazil was a pioneer in this area, UNIDO sponsorship made it possible to promote this initiative in other Latin American countries (Mojica 2004).

The Ibero-American Network of Science and Technology Indicators (RICYT) could be considered as a very good example of "good practice" that could be taken into account as a model for designing a strategy aimed at consolidating FTA capabilities. For just over ten years now, the RICYT has been collecting and publishing data regularly from all the countries of Latin America, as well as Spain and

Portugal, using around fifty comparative indicators (RICYT 1998–2005). Although the information is still discontinuous, for the first time there are also many complete series. These date from 1990 onwards and are updated annually.

RICYT was created in the middle of the 1990s, when it had become apparent that the scientific and technological information available in Latin America was highly unsatisfactory. Most countries lacked reliable and comparable information although some of them had made attempts in the Seventies to gather information on scientific and technological activities. Nevertheless, these attempts faded away in the following decade.

An initial characteristic of the network was that it brought together two diverse sets of actors: on the one hand, national organizations of science and technology, that are simultaneously producers and users of information; and, on the other hand, researchers devoted to studying the relationship between science, technology and society, as well as experts in indicators. This dual participation conditioned both the approach and the agenda. Thus, some concentrated more on designing indicators for policies while others preferred to explore new dimensions. This approach is still present and reflects the type of the actors who participate in the network.

By addressing multiple dimensions, the network proved to be a suitable organizational form for managing a triple heterogeneity:

1. Diversity of involved disciplines
2. Diversity among actors
3. Diversity of capabilities

Diversity makes management more complex, and may prevent the emergence of a culture common and get in the way of efficiency. Nevertheless, it constitutes more of a challenge than a threat to the network and can be seen as an asset offering a whole range of opportunities.

The structural conditions referred to previously involve not only "internal conditions" (training human resources, availability of information and systemic links), but also "external" conditions, such as a new culture in decision-making processes, both at public and private levels. RICYT was able to manage both dimensions of the process, producing better information and promoting its use in the decision-making process in S&T policies. Due to this successful performance, RICYT may be considered as a good example for promoting the application of FTA methods in Latin American countries. In order to do this, it is necessary to take into account the problem of getting people to cooperate, especially key decision makers in science and technology. It is also necessary to reconcile institutional diversities and to answer the specific needs of the countries of the region. In short, it is necessary to manage diversity. But seen in the right way, diversity constitutes a challenge and strength. By bringing together different capabilities, successful initiatives can emerge.

Learning from the previous Latin American experience, FTA exercises should not be aimed at achieving holistic certainties. Rather, FTA could be useful for illuminating concrete scenarios in a general framework of uncertainty, allowing to identify opportunities and to solve problems.

References

Bell, D. (1994): El advenimiento de la sociedad post-industrial, Madrid, Alianza Editorial

Escorsa, P., Maspons, R. and Cruz, E. (2006): Inteligencia competitiva y transferencia de tecnologías: reflexiones para el desarrollo de la relación universidad – empresa, in MEDINA, J. and RINCÓN, G. (eds.)

Herrera, A., et al. (1976): Catástrofe o Nueva Sociedad? El Modelo Mundial Latinoamericano, Ottawa, International Development Research Centre

Herrera, A. et al. (1994): Las Nuevas Tecnologías y el Futuro de América Latina. Riesgo y oportunidad, proyecto Prospectiva Tecnológica para América Latina, Universidade de Campinas

Marí, M. and Callejo, J. (2000): La prospectiva tecnológica y sus métodos, documento de la Secretaría para la Tecnología, la Ciencia y la Innovación Productiva, Argentina

Medina Vásquez, J. (2006): Colombian Program on Technological and Industrial Foresight: Achievements And Perspectives, paper presented at the Second International Seville Seminar on Future-Oriented Technology Analysis (FTA) – September 2006

Medina Vásquez, J. and Rincón Bergman, G. (eds.) (2006): La prospectiva tecnológica e industrial. Contexto, fundamentos y aplicaciones, Bogotá, Conciencias/CAF

Mojica, F. J. (2004): La Prospectiva Tecnológica en el mundo, en América Latina y en los países del Convenio Andrés Bello, Bogotá, CAB

RICYT (1998–2005): El estado de la ciencia. Principales Indicadores de Ciencia y Tecnología Iberoamericanos – Interamericanos, Buenos Aires, RICYT

Varsavsky, O. and Calcagno, E. (1971): América Latina: Modelos Matemáticos, Santiago de Chile, Editorial universitaria, Colección Tiempo Latinoamericano

Part III
FTA Futures

Chapter 11
New Methodological Developments in FTA[1]

F. Scapolo and A.L. Porter

11.1 Introduction

This chapter describes new methodological developments in FTA reported at both the First and Second FTA Seminars. We offer some perspective. This chapter is being prepared as part of the synthesis of knowledge gained especially from the Second Future-oriented Technology Analysis (FTA) Seminar (2006). However, with respect to "new methods," we incorporate considerable material from the First FTA Seminar (2004 EU–US Seminar) because its theme was "new methods". We intend this chapter as an aid to the reader scanning for suitable technology intelligence, forecasting, assessment, roadmapping, or foresight tools. The "foundation papers" for the First FTA Seminar noted several important drivers that imply the need for new methods. Coates et al. (2001) observed that FTA had emerged from an extended dormancy with an upsurge in new forms and incipient new tools in the 1990's. They perceived several potent changes and challenges for FTA:

1. Changes in the nature of "technological change" with increasingly science-based innovation
2. Shift in the prime drivers of technological innovation from the more narrowly technical concerns of Soviet-American Cold War military systems to industrial competitiveness concerns requiring inclusion of socio-economic contextual influences
3. Renewed attention to societal outcomes (and sustainability)
4. Opportunities to exploit electronic information resources to enrich FTA
5. Better capabilities to address complexity in technological innovation

These changes all imply the need to adapt classical tech forecasting methods to address these challenges of informing technology management effectively. Based on later interchanges, we might augment this list of changing challenges to include:

1. Recognition of essential technological innovation process uncertainties that mandate adaptive risk management responses

---

[1] This draws heavily upon the FTA-2006 foundation paper by Rader and Porter (2006). It also excerpts heavily from Porter (2007).

C. Cagnin et al. (eds.) *Future-Oriented Technology Analysis*,
doi:10.1007/978-3-540-68811-2,

2. Interest in discontinuous advances in science and technology, pointing toward radical innovation
3. Suitably engaging multiple stakeholders in participatory FTA processes to assure distributed understanding

11.2 What's New in FTA Methods?

A discussion of "what's new" in terms of methods used for FTA should ground itself by noting "what's old." For many years there have been good treatments of methods in the major FTA activity forms. Not attempting to catalogue these, we note a few (reflecting personal familiarity):

- Technology Intelligence:
  - W.B. Ashton and R.A. Klavens (eds) (1997), *Keeping Abreast of Science and Technology*
- Technology Forecasting:
  - Martino, J. P. (1993), *Technological Forecasting for Decision Making*
  - Porter et al. (1991), *Forecasting and Management of Technology*
- Technology Assessment:
  - Porter et al. (1980a,b), *A Guidebook for Technology Assessment and Impact Analysis*
- Futures Research:
  - Gordon et al. (2003), Futures Research Methods

The great majority of methods used for technological forecasting, assessment, and foresight trace their roots well back. Trend analyses and the Delphi method of obtaining expert opinion have been applied since the 1950s. Delphi began at the Rand Corporation for military analyses (c.f., Linstone and Turoff 1975). Other forms of modelling and scenarios have also been in use for decades.

Since the beginning of the 1990s, with the flourishing of Foresight activities at the national level for science and technology (S&T) priority setting, there have been many efforts in categorising and classifying the various methods and techniques used for FTA. A number of handbooks and guidelines have come forth on how to do Foresight. All these have, to some extent, material on methods and techniques that can be used for Foresight, not only at the national level, but also at regional and sectoral levels – and not only for S&T priority setting. These works have been very valuable and useful. In most cases they consider methods starting from an analysis of the requirements for implementation in terms of data and inputs, how the data (various form i.e. expert judgement to number) are treated by the various methods, what are the outcomes, and how the outcomes can be used.

These references underline that the selection of a methodological approach for an FTA study depends on a number of variables. The context of the exercise is very important, of course, but there are other variables that might affect the selection of methods. These include: the time-horizon of an exercise, its duration, the methodological competence of the people that are managing the exercise, and last, but not least, availability of resources. The implementation of some techniques is more costly than others, especially if the study relies on the involvement of experts.

Despite the wealth of FTA studies and development of specialised literature addressing FTA processes, there has not been a systematic effort to address new methods and techniques for FTA studies. This chapter attempts to scan such methods. It attends to the novel application of existing methods, the combination of techniques within a single FTA exercise, and any new methodological development.

11.3 FTA Methods

The working group which prepared for the first Seville seminar (Technology Future Analysis Methods Working Group, 2004) proposed a typology consisting of 51 methods arranged in nine "families," modified slightly by Porter (forthcoming) into some 50 or so methods in 13 families (Table 11.1).

While a standard recipe for successful FTA analyses is not sensible, a great deal of experience with such activities exists from which to draw lessons. In many cases we can state which methods have worked well toward attaining which goals, given particular content emphases, contingent upon processes involving certain participants and clients. We seek to identify more general recommendations related to FTA activities ("do's" and "don'ts").

In order to develop such guidelines for the selection of methods well-suited for the specific circumstances of an FTA activity, an important step is to analyse existing experience through case studies. We seek to pinpoint critical factors in matching goals, motivations and clients with methods and other dimensions, such as approach and organisation of FTA processes.

The methodological families in Table 11.1 are roughly ordered from descriptive toward prescriptive. Many emphasize gathering and portraying data. Creativity approaches intend to broaden our consideration; to prod us "out of the box," helping to avoid falling into the "Zeitgeist" (conventional wisdom) trap. Monitoring and intelligence methods draw in and profile available information. Descriptive methods and matrices massage that information to facilitate interpretation. Descriptive statistics are embedded in several of the approaches (e.g., trend analyses), but FTA is not overly reliant on extensive statistical manipulations. Trend Analyses – historical time series data and their projection into the future – are basic FTA tools. Expert Opinion sometimes stands alone; even better, it can be combined with empirical approaches to help integrate and interpret. Visual methods, such as science theatres or short films, have proved useful when exercises seek to involve lay persons.

Table 11.1 Future-oriented technology analysis methods

Methods families	Sample methods
Creativity approaches	TRIZ, Future workshops, visioning
Monitoring & intelligence	Technology watch, tech mining[2]
Descriptive	Bibliometrics, impact checklists, State of the future index, multiple perspectives assessment
Matrices	Analogies, morphological analysis, cross-impact analyses
Statistical analyses	Risk analysis, correlations
Trend analyses	Growth curve modelling, leading indicators, envelope curves, long wave models
Expert opinion	Survey, delphi, focus groups, participatory approaches
Modelling & simulation	Innovation systems descriptions, Complex adaptive systems modelling, chaotic regimes modelling, technology diffusion or substitution analyses, input-output modelling, agent-based modelling
Logical/ causal analyses	Requirements analysis, institutional analyses, stakeholder analyses, social impact assessment, mitigation strategizing, sustainability analyses, action analyses (policy assessment), relevance trees, futures wheel
Roadmapping	Backcasting, technology/product roadmapping, science mapping, multi-path mapping (Robinson and Propp 2006)
Scenarios	Scenario management, quantitatively based scenarios, different emphases,[3] science theatres, video (Steyaert et al. 2006, Decker and Ladikas 2004)
Valuing/decision-aiding/economic analyses	Cost-benefit analysis (CBA), SWOT and scorecard analyses,[4] analytical hierarchy process (AHP), data envelopment analysis (DEA), multicriteria decision analyses
Combinations	Scenario-simulation (gaming), trend impact analysis

[2] Porter and Cunningham (2005) describe approaches to exploit ST&I information resources; van de Lei and Cunningham (2006) extend to "web mining."

[3] Punie et al. (2006) argue for the value of "dark scenarios" to direct stronger attention to safeguard issues.

[4] Sripaipan (2006) illustrates for APEC foresight work.

The remaining methods families entail more manipulation of the data. Modelling & Simulation cover a wide gamut – from qualitative modelling ("boxes and arrows" pointing toward the Logical/Causal Analyses) to intricate, quantified, computer modelling. Logical/Causal Analyses trace "if – then" relationships to help draw implications. Roadmapping weaves these into future progressions, particularly to inform S&T planning. Scenarios combine multiple elements to convey alternative futures. These and the Valuing/Decision-aiding/Economic analyses point toward assessing policy/ action options. Combinations are just that – interesting ways to integrate different tools to gain perspective for better foresight.

Let's take six base FTA methods (the 12 more detailed family types of Table 11.1 map pretty clearly to these):

- Monitoring
- Creativity methods
- Trend analyses
- Simulation and modelling
- Expert opinion
- Scenarios

Many of the new methods reflect variations on these base methods. We take the liberty of using these six to organize consideration of a number of novel approaches. In some cases, categorization is pretty arbitrary. The intent is rather to introduce these ideas and encourage readers to explore particular ones of interest further, beginning with the FTA Proceedings (IPTS 2004).

Monitoring. This family of methods includes many variants, such as environmental scanning, technology alerts, and competitive technical intelligence. All share the basic approach of perusing a body of information (often, but certainly not exclusively, R&D publication and patent abstract sets), digesting pertinent messages for an organization's interests, and interpreting the implications. What's new here consists, first, of the increasing availability of S&T information in handy, electronic form. As Bill Gates put it, we now have incredible "information at our fingertips". This includes the major R&D databases, such as Web of Knowledge (including Science Citation Index and Social Science Citation Index), EI Village (with access to the two leading engineering databases, EI Compendex and INSPEC), MEDLINE (covering biomedical research), and the world patent databases (e.g., Derwent World Patent Index, Micropatent), and many others (e.g., RaDiUS covering U.S. Federal research projects or NSF's Awards Database). Beyond these are business, popular press, venture capital, standards, and other contextual information resources. And, of course, the Internet offers access to a wealth of individual research-oriented websites plus fascinating compilations. So, why use the databases? They scan multiple sources (e.g., thousands of journals and/or conferences), filter, format, and index the raw information. They also provide search & retrieval capabilities otherwise lacking. That is, in a minute or so, one can locate and download thousands of well-structured abstract records. For instance, if you were analyzing a particular agent-organ-disease combination, a search in MEDLINE's collection of 12,000,000 or so world bio-medical article abstracts could provide excellent, first-order coverage of years of research results.

Established bibliometric approaches track R&D activity patterns. New methods incorporate "text mining" to discern trends and relationships. These allow comprehension of entire research domains via "research profiling" (Porter et al. 2002). More generally they allow the analyst to answer "who, what, where, and when" questions about research activity – i.e., to generate technical intelligence (c.f., Porter and Cunningham 2005). Boyack and Rahal (2004) addressed the vital step beyond analysis – illustrating interactive information visualization possibilities.

Creativity methods. This diverse family of methods seeks fresh ideas for technologies, their fusion, and new applications. TRIZ is an important approach, deriving from Russian patent analyses, to draw upon analogous problem solutions from other domains. Using typologies of technical changes and challenges, one abstracts the problem at hand so as to recognize analogous solutions that may offer new perspectives to try out. TRIZ applications are reaching into "Management of Technology" (MOT), not just technical issues (c.f., Mohrle 2000).

A standby creativity approach is brainstorming. Van Notten (2004) presented a new approach of staggered brainwriting. An important way to enrich new product development is to move from "technology push" to "societal pull" considerations. Green (2004) described how a major company effectively takes into account social drivers in the design process in the form of "cultural creatives."

Trend analyses. Trends (time series data) lend themselves to growth modelling and extrapolation (trend projection). The S-shaped growth curves (e.g., logistic growth patterns showing an essentially exponential growth phase tapering into an asymptotic approach to a limit) are prevalent in tech forecasting. Technology or product families often witness successions of such S-curves.

Linstone (2004) and others point to the increasing extent of chaotic, transition regimes interspersed with smooth growth regimes. On the one hand, these imply intrinsic limits to forecasting – this suggests that rapid analyses and adaptive MOT are increasingly necessary. On the other, it lends importance to new methods to address the chaotic regimes (see next segment).

Simulation and modelling This family of methods encompasses a wide range of approaches. The FTA Seminar treated both quantitative and qualitative modelling. Indeed, modelling in one form or another is incorporated into a good number of the new methods noted under the other headings as well.

Complex adaptive systems (CAS) modelling reaches out, for instance, to use genetic algorithms in treating technological innovation processes. Devezas (2004) presented evolutionary process modelling, using interactive growth modelling (Lotka-Volterra). He highlights the theoretical affinities among biological, cultural, and technological change processes.

De Jouvenel (2004) presented a modern cross-impact approach. By crossing technological "push" attributes with social "pull" in the form of values, needs, and objectives, one can adjust a developing technology in its formative stage. Looking downstream, Pals et al. (2004) described how to operationalize consideration of the behavioural factors attendant to adoption of a technological innovation. They have a matrix of 26 product characteristics crossed against 14 personality traits of target groups to assess fit.

Interactive simulation, or gaming, offers a fresh tool for FTA. Gaming approaches are reaching beyond war games to address competitive interplay among technologies and products. Salo et al. (2004) introduced gaming in the context of climate change.

"Lock-in" was noted as an essential consideration in anticipating technology "winners". This is not a new method per se, but rather a feature to consider in modelling innovation processes. The classic example of VHS vs. Beta VHR systems is well known. VHS, by gaining the initial success locked out Beta to a large extent. Another example is the long run of semiconductor innovations building on the silicon platform. So much capital investment and experiential learning make it very difficult for alternative technologies (in this case, gallium arsenide semiconductors) to catch on. Fleischer et al. (2004) presented lock-in as a factor in considering alternative development pathways and likelihoods of successful innovation for nanotechnologies.

Daim et al. (2006) present a study that makes use of bibliometrics and patent analysis to forecast emerging technologies where there is a lack of availability of historical data. They describe the forecasts for three emerging technology areas by integrating the use of bibliometrics and patent analysis into well-known technology forecasting tools such as scenario planning, growth curves and analogies. System dynamics is also used to be able to model the dynamic ecosystem of the technologies and their diffusion.

"Roadmapping" (i.e., planning stepwise through a series of interrelated technologies and, often, products) might well be considered as a separate form of FTA. But, let's mention it here. Multiple papers at the 2004 FTA Seminar addressed variations on technology roadmapping (c.f., Price et al. 2004). De Laat (2004) provided a meta-analysis of 80 roadmapping exercises to identify conditions for success. Moving toward new versions, Green (2004) described how roadmapping can be adapted to a product design setting. Fiedeler et al. (2004) discussed a "cross-form" application – namely, use of roadmapping for technology assessment (specifically, impact assessment for nanotechnology). Lizaso and Reger (2004) crossed methods in relating scenario-based roadmapping.

Oner and Saritas (2005) proposed two methodologies, namely Integrated Management Model (IMM) and Roadmapping, in order to overcome challenges introduced by the multidimensional characteristics and complex nature of foresight studies. Based on systemic approach, IMM offers a useful way of developing long-term normative policies and strategies and their transformations into actions by considering necessary changes in organizational structures and behaviours. In addition, roadmapping is used to capture, manipulate and manage information to decrease complexity in the foresight by constructing roadmaps. In the paper, IMM and roadmapping are employed first to analyze UK foresight results and then to develop a new methodology to formulate Delphi events and scenarios for the successful implementation of foresight. This study also promotes the integrated use of foresight techniques such as scenarios and Delphi rather than one for another.

Recent FTA studies show that one of the aims of the application of the various methods and tools is trying to reduce uncertainties such as to forecast accurately the future of a technological area. The goal is to identify and apply the most adequate method to develop useful knowledge on breakthrough points, or degrees of market penetration of a technological area. Cheng et al. (2007) applied the fuzzy analytic hierarchy process (FAHP) method (i.e. one of the most popular multiple criteria decision-making method) to evaluate and select the technology forecasting methods in the field of new materials. The fuzzy set theory could resemble human reasoning in use of approximate information and uncertainty to generate decisions. It was specifically designed to mathematically represent uncertainty and vagueness and provide formalized tools for dealing with the imprecision intrinsic to many problems. In their study they have identified the critical evaluation criteria to evaluate the technology forecasting methods for development of new materials.

Expert opinion. The inclusion of some form of expert knowledge is virtually a given in FTA. Scapolo and Miles (2006) compared alternative ways to gather expert inputs. In the conclusions the authors could not reach any firm conclusion on the role of consulting experts and on how to systematically feed the knowledge generated through the application of various tools as one contribution to the policy making processes. They suggest that further research and work in this respect would be valuable. Eerola et al. (2004) discussed another methodological combination – expert opinion with formal analyses.

Participatory issues are prominent in technology foresight. Berloznik (IPTS 2004) offered another form of combined methodology. He described the use of consensus-building games to enhance participation.

The review of recent literature continues to show that the Delphi method is still one of the most (or the most) known methods to elicit expert opinion in relation to FTA studies.

Some interesting recent applications of the Delphi method explore a variety of contexts with the support of advanced information technology to overcome the use of sequential rounds that delays the efficiency of the process (Gordon and Pease 2006). "Real Time Delphi" also has the advantage of providing almost real time results. The implementation requires a single round and functions via Internet or other electronic network. The respondents can join in at different stages and are presented with preliminary results of the groups (i.e., the median) and can reply to the questionnaire. Each respondent views his/her own earlier response upon returning to the process and can revise it, particularly in case it falls outside the interquartile range. As respondents continue to watch their input form (or later on a return visit), they also see new averages, distributions, and reasons given by other panelists for their positions. This information appears whenever new inputs are received from other participants.

The applications of this method include forecasting, foresight, and policy studies, in any problem for which the synthesis of expert opinion is necessary or desired.

Scenarios. This family of methods is widely practiced in many forms (Mietzner and Reger 2004). The literature is rich in variations. Again, the theme of methodological cross-fertilization was well-represented. Fontela and Rueda-Cantuche (2004) devised scenarios based on probabilistic cross-impact analyses. Barré and David (2004) also worked with a quantitative, input/output-based scenario approach. Elsewhere, Lempert (2002) has written about agent-based modelling that can yield a million alternative scenarios with ways to reduce these to effective policy options.

An often neglected consideration regarding methods is their suitability to generate results that FTA users find compelling. Gaßner and Steinmüller (2004) showed the potential benefits of a story-telling scenario option.

Another intersection came between one of the eight changes/challenges – namely the importance of discontinuous advances in S&T – and scenarios. Van Notten (2004) compared alternative approaches for scenario development concentrated on discontinuous change. Postma and Liebl (2005) address some of the current drawbacks of the conventional scenario method and indicate possible avenues for methodological adaptations.

More recently a literature review on the use and advancements on scenario method, underlines that the use of scenario planning remains one of the most applied methods in FTA work. There have been, over the years, various attempts to reflect and improve upon the effectiveness of the method. A number of studies look at how scenario planning can be improved to deal with uncertainties. Scenario planning is widely accepted as one of the key contemporary approaches to help organisations explore environmental uncertainty, helping to cope with change in their business environment. For example, Burt (2007) focuses on the integration of Christensen's theory of disruption with the scenario methodology to develop a framework that provides an understanding of the underlying systemic conditions that create disruption and/or discontinuity. Burt improves understanding of the methodology and its practice by developing a theoretical framework to help scenario planners identify and understand environmental conditions that would result in disruption and/or discontinuity.

Another application of scenarios analysis was done in combination with trend analysis on market exploration for breakthrough technologies (Ortt et al. 2007). This study combined market research and future research by using the outputs of market research as inputs for future research. Market research is generally not an appropriate approach to indicate discontinuous market changes such as when new technologies enter the market or when events disturb the market. Trend-analysis can indicate how needs have been experienced in the course of time and which aspects in the environment have an effect on the needs. If various aspects will stimulate or block the need in the future then a discontinuous change in the way this need is felt. When trends become uncertain, then scenario analysis can be applied to explore the potential consequences. Used in combination, market research and futures research are powerful for understanding new market conditions and guiding product development and marketing, particularly when explorations are needed most but are most difficult to obtain.

In addition, scenario analyses have been combined with other methods, especially with modelling techniques. One study on the future of broadband technologies (Wang and Lan 2007) combined scenario analysis and technological substitution modelling (a quantitative method) to explore new generation technological development. Scenario analysis incorporates uncertainty but fails to provide sustained quantitative forecasts. The technological substitution model offers quantitative forecasts but may neglect key influences for the possible future. The combination of the two methods offered six strengths (Wang and Lan 2007). For example the combined qualitative and quantitative approach allowed considering the dynamic changes in the business environment, it provided a prediction of the annual developments, and helped to forecast the replacement of old technologies. Decision making in the face of such a technological transition is thereby better informed.

11.4 Conclusions

In Chap. 3, Box 3.1 highlights "Ten Commandments" to fit methods to FTA types to achieve goals. Some elements of the box presented in Chap. 3 are valid and relevant for the future of FTA methods. One in particular is of crucial importance: RESEARCH on FTA method needs support. For some of the methods the question of reliability of data treatment and generation of outcomes (e.g. cross-impact analysis and modelling) still remains a serious concern. Indeed, one way to enhance knowledge about FTA methods is to continue applying (thereby improving) existing methods, because developing new ones requires considerable time and resources (but is also desirable). We have observed a trend in recent FTA studies that we encourage, which is the use of multiple methods. We especially consider that the combination of qualitative and quantitative methods (i.e. empirical and expert opinion) can provide more robust results and information, and, in this way, offer decision makers grounded information on which they can base their strategies and decisions.

Attempting a bit of foresight for FTA as such, three themes for FTA methods development merit consideration. First, the increasingly widespread availability of data of all sorts is not likely to abate, making advanced tools that help process, search, mine, organize, display and interpret electronic information resources essential (Porter and Cunningham 2005). Second, the need for better methods of extracting, organizing, comparing and combining a wide variety of human judgments warrants attention. Taking a vast array of expressed interests and opinions into account seems to be a continuing driver to improve FTA studies. It is also essential to underline the scientific quality and validity of such participatory approaches which are a result of new governance concepts. Third, proliferation of rapid communication tools (e.g., internet-based) will permit vast numbers of "anywhere" participants in FTA studies. Electronic voting processes could contribute to FTA processes. Networking and collaboration tools should facilitate contributions from diverse stakeholders. A key in going forward is to cumulate evaluative knowledge as to the effectiveness of given FTA exercises and the roles played by particular methods therein. We urge sharing of experiences and point toward doing so through the EFMN network (Keenan et al. 2006 – www.efmn.eu) and the COST Action A22 "Advancing Foresight Methodology" endeavor (Fuller et al. 2006).

What initiatives could most affect development of FTA? We extrapolate beyond this paper's content to offer suggestions:

- Assessment of FTA studies to generate "lessons learned" as to what methods best suit which types of studies is vital.
- Community-building is needed. We need to actively interchange perspectives and practices across the range of FTA types. In particular, this should engage analysts and scholars; it should bring together public and private sector interests.
- Training and education need attention. To the best of our knowledge, university educational attention to FTA has dwindled badly. FTA-oriented courses could foster interest in the field, scholarship, and able future practitioners.
- Research on FTA has likewise waned horribly. In the 1970s in the US, a small, but active, National Science Foundation program on Technology Assessment stimulated tremendous academic engagement and helped generate a professional community. Given US disinterest, perhaps the EU Research Directorate General would consider support for methodological development, FTA evaluation, and explorations into new approaches.

References

Ashton, W.B.,Klavens, R.A. (eds) (1997) Keeping Abreast of Science and Technology, Battelle Press, Columbus, Ohio, USA

Barré R., David B. (2004) 'Participative & Coherent Scenario building: An Input/ Output Balance model; The Case of the French National FutuRIS Operation', paper presented at the First EU–US Seville Seminar on Future-Oriented Technology Analysis (FTA) – May 2004

Boyack K.W., Rahal N. (2004) Evaluation of Laboratory Directed Research and Development (LDRD) Investment Areas at Sandia, paper presented at the First EU–US Seville Seminar on Future-Oriented Technology Analysis (FTA) – May 2004

Burt, G. (2007) Why are We Surprised at Surprises? Integrating Disruption Theory and System Analysis with the Scenario Methodology to Help Identify Disruptions and Discontinuities in Technological Forecasting and Social Change, Vol. 74, No. 6, pp. 731–749

Cheng, A.-C., Chen, C.-J., Chen, C.-Y. (2008) A fuzzy multiple criteria comparison of technology forecasting methods for predicting the new materials development in Technological Forecasting and Social Change, available on Science Direct, Vol. 75, No. 1, pp. 131–141

Coates, V., Faroque, M., Klavins, R., Lapid, K., Linstone, H.A., Pistorius, C., Porter, A.L. (2001) On The Future of Technological Forecasting, Technological Forecasting and Social Change, Vol .67, No. 1, pp. 1–17

Daim, T.U., Rueda, G., Martin, H., Gerdsri, P. (2006) Forecasting emerging technologies: Use of bibliometrics and patent analysis' in Technological Forecasting and Social Change, Vol. 73, No. 8, pp. 981–1012

Decker, M., Ladikas, M. (eds) (2004) Bridges between Science, Society and Policy. Technology Assessment – Methods and Impacts, Springer, Berlin Heidelberg New York

De Jouvenel H. (2004) Science and Society: Towards an Integrative Approach in Foresight, paper presented at the First EU–US Seville Seminar on Future-Oriented Technology Analysis (FTA) – May 2004

De Laat B. (2004) Conditions for Effectiveness in Roadmapping: a Cross-sectional Analysis of 80 Different Exercises, paper presented at the First EU–US Seville Seminar on Future-Oriented Technology Analysis (FTA) – May 2004

Devezas T.C. (2004) Evolutionary Theory of Technological Change: State-of-the-art and New Approaches, paper presented at the First EU–US Seville Seminar on Future-Oriented Technology Analysis (FTA) – May 2004

Eerola A., Loikkanen T., Joergensen B.H., Andersen P.D., Eriksson E.A. (2004) 'Nordic H_2 Energy Foresight – Complementary contribution of expert views and formal analyses', paper presented at the First EU–US Seville Seminar on Future-Oriented Technology Analysis (FTA) – May 2004

Fiedeler U., Fleischer T., Decker M. (2004) Roadmapping as TA-Tool: Pre-requisites and potential benefits for assessing Nanotechnology, paper presented at the First EU–US Seville Seminar on Future-Oriented Technology Analysis (FTA) – May 2004

Fleischer T., Decker M., Fiedeler U. (2004) Assessing Emerging Technologies – Methodical Challenges and the Case of Nanotechnologies, paper presented at the First EU–US Seville Seminar on Future-Oriented Technology Analysis (FTA) – May 2004

Fontela E., Rueda-Cantuche J.M. (2004) Linking Cross-Impact Probabilistic Scenarios to Input-Output Models, paper presented at the First EU–US Seville Seminar on Future-Oriented Technology Analysis (FTA) – May 2004

Fuller, T., De Smedt, P., and Rothman, D.S. (2006) Advancing Foresight Methodology Through Networked Conversation, FTA 2006 Seminar (Future-oriented Technology Analysis), Institute for Prospective Technological Studies, Seville, 28–29 September

Gaßner R., Steinmüller K. (2004) Scenarios that Tell a Story; Normative Narrative Scenarios – An Efficient Tool for Participative Innovation-oriented Foresight, paper presented at the First EU–US Seville Seminar on Future-Oriented Technology Analysis (FTA) – May 2004

Gordon, T.J., Glenn, J.C., Peter Bishop, P. (2003) From a CD ROM on Futures Research Methods, published by the Millennium Project of the American Council for the United Nations University [http://www.acunu.org/millennium/FRM v2.0]

Gordon, T., Pease, A. (2006) RT Delphi: An Efficient, "Round-less", Almost Real Time Delphi Method. In Technological Forecasting and Social Change, Vol. 73, pp. 321–333

Green J. (2004) Co-evolving Technology Roadmaps – Roadmaps as living tools, paper presented at the First EU–US Seville Seminar on Future-Oriented Technology Analysis (FTA) – May 2004

IPTS (Institute for Prospective Technological Studies) (2004) New Horizons and Challenges for Future-oriented Technology Analyses (FTA), Proceedings of the EU–US Scientific Seminar: New Technology Foresight, Forecasting and Assessment Methods, JRC-IPTS – Seville, 13–14 May 2004

Keenan, M., Butter, M., Sainz de la Fuenta, G.,and Popper, R. (2006) Mapping Foresight in Europe and Other Regions of the World: The 2006 Annual Mapping Report of the EFMN, European Foresight Monitoring Network [www.efmn.eu]

Lempert, R.J. (2002) Agent-Based Modeling as Organizational and Public Policy Simulators, Proceedings of the National Academy of Sciences (PNAS), Vol. 99, Sup. 3, 7195–7196

Linstone H.A. (2004) New Drivers and Directions: Challenges for FTA, paper presented at the First EU–US Seville Seminar on Future-Oriented Technology Analysis (FTA) – May 2004

Linstone, H.A., Turoff, M. (eds.) (1975) The Delphi Method: Techniques and Applications, Addison-Wesley, Reading, MA, USA

Lizaso F., Reger G. (2004) Scenario-Based Roadmapping – A Conceptual View, paper presented at the First EU–US Seville Seminar on Future-Oriented Technology Analysis (FTA) – May 2004

Martino, J. P. (1993). Technological Forecasting for Decision Making, third edition, McGraw-Hill, New York.

Mietzner, D., Reger, G. (2004) Scenario Approaches – History, Differences, Advantages, and Disadvantages, paper presented at the First EU–US Seville Seminar on Future-Oriented Technology Analysis (FTA) – May 2004

Mohrle, M.G. (2000). TRIZ-Based Competitor Analyses, Ninth International Conference on Management of Technology, International Association for Management of Technology (IAMOT), Miami, February 20–25

Oner, M.A, Saritas, O. (2005) A Systems Approach to Policy Analysis and Development Planning: Construction Sector in the Turkish 5-Year Development Plans in Technological Forecasting and Social Change, Vol. 72, No. 7, pp. 886–911

Ortt, J.R, Langley, D.J., Pals, N. (2007) Exploring the Market for Breakthrough Technologies in Technological Forecasting and Social Change, Vol. 74, No. 9, pp.1788–1804

Pals N., Langley D.J., Ortt J.R. (2004) Evolution of Behaviour Predicting the Adoption of Innovations, paper presented at the First EU–US Seville Seminar on Future-Oriented Technology Analysis (FTA) – May 2004

Porter, A.L. (forthcoming): Technology Foresight: Types and Methods, International Journal of Technology Management, forthcoming

Porter A.L. (2007) New Methods for Technology Futures Analyses, in M.H. Sherif and T.M. Khalil (eds.), Management of Technology: New Directions in Technology Management, Elsevier, 453–464

Porter, A.L., and Cunningham, S.W. (2005) Tech Mining: Exploiting New Technologies for Competitive Advantage, Wiley, New York

Porter, A.L., Rossini, G.A., Carpenter, S.R., Roper, A.T. (1980a) A Guidebook for Technology Assessment and Impact Analysis, North-Holland, New York

Porter, A.L., Rossini, F.A., Carpenter, S.R. Roper, A.T. (1980b) A Guidebook for Technology Assessment and Impact Analysis, North-Holland, New York

Porter, A.L., Roper, A.T., Mason, T.W., Rossini, F.A., Banks, J. (1991) Forecasting and Management of Technology, Wiley, New York

Porter, A.L., Kongthon, A., Lu, J-C. (2002) Research Profiling: Improving the Literature Review, Scientometrics, Vol. 53, pp. 351–370

Postma, T. J.B.M and F. Liebl (2005) How to Improve Scenario Analysis as a Strategic Management Tool?' in Technological Forecasting and Social Change, Vol. 72, No. 2, pp.161–173

Price, S., Conway, P., Palmer, P., Summers, R. (2004) Technology Roadmapping – a New Perspective, paper presented at the First EU–US Seville Seminar on Future-Oriented Technology Analysis (FTA) – May 2004

Punie, Y., Maghiros, I., and Delaitre, S. (2006) Dark Scenarios as a Constructive Tool for Future-Oriented Technology Analysis: Safeguards in a World of Ambient Intelligence (SWAMI), FTA 2006 Seminar (Future-oriented Technology Analysis), Institute for Prospective Technological Studies, Seville, 28–29 September

Rader M., Porter A. (2006) Fitting Future-Oriented Technology Analysis Methods to Study Types, FTA 2006 Seminar (Future-oriented Technology Analysis), Institute for Prospective Technological Studies, Seville, 28–29 September

Robinson, D.K.R., Propp, T. (2006) Multi-Path Mapping as Strategic Intelligence for Reflexive Alighment in Emerging S&T, FTA 2006 Seminar (Future-Oriented Technology Analysis), Institute for Prospective Technological Studies, Seville, 28–29 September

Salo A., Gustafsson R., Kaakkolammi I., Gustafsson T. (2004) Participatory Development of Future-Oriented Policy Configurations for Climate Change and Mobile Gaming, paper presented at the First EU–US Seville Seminar on Future-Oriented Technology Analysis (FTA) – May 2004

Scapolo, F, Miles, I. (2006) Eliciting Experts' Knowledge: A Comparison of Two Methods in Technological Forecasting and Social Change, Vol. 73, No. 6, pp. 679–704

Sripaipan, C. (2006) Foresight in a Non-Profit Organisation: A Case of the Technology Promotion Association (Thailand-Japan), FTA 2006 Seminar (Future-oriented Technology Analysis), Institute for Prospective Technological Studies, Seville, 28–29 September

Steyaert, S., Eggermont, S., Vandebosch, H. (2006) Towards the Desired Future of the Elderly and ICT: Policy Recommendations Based on a Dialogue with Senior Citizens. FTA 2006 Seminar

(Future-Oriented Technology Analysis), Institute for Prospective Technological Studies, Seville, 28–29 September

Technology Futures Analysis Methods Working Group (2004) [Porter, A.L., Ashton, B., Clar, G., Coates, J.F., Cuhls, K., Cunningham, S.W., Ducatel, K., van der Duin, P., Georghiou, L., Gordon, T., Linstone, H., Marchau, V., Massari, G., Miles, I., Mogee, M., Salo, A., Scapolo, F., Smits, R., Thissen, W.], Technology Futures Analysis: Toward Integration of the Field and New Methods, in Technological Forecasting and Social Change, Vol. 71, pp. 287–303

van de Lei, T.E., and Cunningham, S.W. (2006) Use of the Internet for Future-Oriented Technology Analysis, FTA 2006 Seminar (Future-Oriented Technology Analysis), Institute for Prospective Technological Studies, Seville, 28–29 September

van Notten P. (2004) A Shocking Experiment: Alternative Approaches for Discontinuity-Oriented Scenario Development, paper presented at the First EU–US Seville Seminar on Future-Oriented Technology Analysis (FTA) – May 2004

Wang, M-Y., Lan, W-T. (2007) Combined Forecast Process: Combining Scenario Analysis with the Technological Substitution Model, in Technological Forecasting and Social Change, Vol. 74, No. 3, pp. 357–378

Chapter 12
Future-Oriented Technology Analysis: Future Directions

M. Keenan, R. Barré and C. Cagnin

As the chapters in this volume have demonstrated, FTA cannot be defined by a single methodology, nor by a single goal; indeed, it encompasses the tradition of a variety of schools of thought, each one embodying a stream of theory and practice referring more or less explicitly to an understanding of what is relevant and useful. The diversity of contexts of application leads understandably to a diversity of focus, methods and outcomes of FTA exercises. Nevertheless, on reviewing the different contributions from which this book is built, it appears that the term 'FTA' does refer to a number of common elements, beyond the differences of context of application and the variety of mode of expression linked to the various backgrounds among the authors. These deep-rooted common elements are the basis for a shared understanding of FTA, as follows:

- FTA is an agenda-setting process aimed at providing anticipatory intelligence as a basis for decision making. It is the set of activities dealing with statements about long term dynamics of technology in society, either to produce such statements or to perform criticism of existing ones. As a consequence, FTA processes initiate collective learning and vision building which impact the complex interplay of factors governing innovation trajectories.
- In FTA, technical change is considered a socially embedded driver, which is key for societal evolution, developing along a time axis, in the form of 'trajectories', hence the need for a longer term horizon of analysis and a broadening of the set of parameters to take into consideration. In this sense, longer term, systemic analysis constitutes a key characteristic of FTA, which explicitly deals with complex socio-technical systems and science-society relationships.
- FTA allows for the construction of common visions and produces issue-specific knowledge through a process of dialogue, creating joint learning spaces between users and producers of innovation, knowledge integration and a shared sense of commitment. In this sense, it is an infrastructure of distributed intelligence – enabling the system to better address future challenges.
- Not surprisingly, FTA has relevance in all human activities where there are collective stakes, as shown in the contributions to this book: it is used in industry, in higher education, in public policy in a variety of countries, both in the innovation field, but more generally regarding socio-economic development. Thus,

C. Cagnin et al. (eds.) *Future-Oriented Technology Analysis*,
doi:10.1007/978-3-540-68811-2,

different FTA exercises can have large differences in scope (geographic scale and time horizon), relationship to decision making, the extent of participation and even the purpose of the analysis.
- At the same time, however, all FTA processes share the following types of outputs: structured and validated information on longer term social and economic developments, identification of solutions to complex problem areas, and defined priority areas.
- They also share criteria for assessing the quality of the FTA processes:
  - Rigour, standing, interest and credibility of the conjectures made
  - Diversity of the actors participating in the debates and their effective access to the forum
  - Impacts in terms of learning effects
  - Impacts in terms of strategy formulation for action by the actors of the system
- Finally, the following criteria would seem to be important for assessing the impacts of FTA:
  - Credibility, depending on the internal validity and analytical rigour of the conjectures made
  - Quality of the conjectures, linked to their creativity and the extent to which they transcend existing beliefs and innovation patterns, i.e. the extent to which they are 'disruptive' and lead to learning effects
  - Social robustness, based on the fact that all viewpoints have been negotiated, both in framing the question and in gathering the evidence
  - Clarity of purpose, meaning that a clear linkage exists between the FTA process and the formal decision-making process
  - Legitimacy resulting from the existence of cooperative strategies for knowledge production

All in all, the major result of this collective effort, undertaken by a variety of professionals, from many countries, and with various backgrounds, is the empirical demonstration of the existence of a Future-oriented Technology Analysis academic and professional community, actively involved in a collective appraisal of its activities to better contribute to a vital social imperative: a better accounting of the long term in decision-making processes, allowing for collective identification and debate of alternative strategies.

In this sense, an important goal of the FTA community is to address the imperative of improving the two-way linkage between knowledge and the building of a "common world". At the same time, it is important to grasp the community's readiness to address global issues – and to building governance at a global level. To achieve such an influence and to be able to deal with the increasing uncertainty and complexity of global issues, the objective must be to establish practical pathways, with clear milestones, outlining the roles and actions that the FTA community would need to undertake in the years to follow. In this context, issues such as practitioners' ethics and how to increase stakeholders' awareness and use of FTA as a policy and decision making instrument will need to be further tackled, for example, by delineating possible synergies in FTA education and training across Europe and beyond. These, and other practical issues, are discussed in more detail below, drawing upon the ideas of participants of the FTA seminars organized to date. Ideas are organized under the following five headings:

- Building capacity in the FTA community
- Building community links
- Raising awareness of FTA among potential users
- Preparedness to address global problems
- Evaluation, monitoring and quality control

12.1 Building Capacity in the FTA Community

All epistemic communities must renew themselves if they are to be sustainable. They must also be useful to society and sufficiently flexible to apply their know-how to new uses in new settings. In the context of the FTA community, this translates into the following needs:

1. Consolidation: through knowledge exchange and the strengthening of ties between FTA practitioners, with a view to the exchange of good practice and the diffusion of new developments in the field.
2. Expansion: through the diffusion of FTA to a wider array of policy and business settings.
3. Renewal: through the provision of training courses at all levels, but particularly at the University level.

An important objective of the FTA seminar series has been to foster a stronger community of future-oriented technology analysts that can bring the latest practices and theories to bear in their work. This has involved exchanging concepts and practices between practitioners in different communities, particularly technology foresight and technology assessment, and, to a lesser extent, technology forecasting. It has also seen the participation of FTA practitioners and users from different worlds, including policy making (especially in S&T, but also in other areas such as environment and energy), corporate strategy and futures, and (to a lesser extent) education and social planning. Furthermore, the seminars have successfully sought to reach out to other parts of the world – beyond the normally Eurocentric focus of the European Commission – in an effort to support a global dialogue.

These are surely positive developments, but they need to be further built upon. Of course, a strong community of practitioners and theoreticians cannot be built overnight, although this is not the real challenge, since the nascent FTA community already has some strong foundations, nay, pillars, on which to build, with existing communities in technology assessment, technology foresight, technological forecasting, and futures studies already having their own journals and societies/networks. Further consolidation efforts are underway in the shape of the EC-funded project on mapping foresight exercises (through the European Foresight Monitoring Network)[1] and the STOA initiatives around technology assessment (summarised in Chap. 3). Moreover, privately-funded infrastructures, such as the Shaping Tomorrow portal[2] and the Prospective-Foresight Network[3], are a sign of a burgeoning demand for future-oriented information.

There is still some way to go to further the provision of training in FTA. Training has been a long-term concern of the futures studies field, which has always struggled to find its rightful place in the academy. The FTA community could learn from this experience, though is perhaps at a slightly different starting point, given its evolution from science, technology and innovation studies. One positive development is the recent growth in the number of doctoral candidates working on FTA, evidenced by the participation of so many young researchers and practitioners in the FTA seminars. Besides the academy, there is surely space for professional education in FTA, with the JRC-IPTS and UNIDO (among others) being active in organising short courses on technology foresight for decision makers. These should be expanded and emulated by other training providers.

12.2 Building Community Links

As we have already seen, the primary objective of the FTA seminar series has been to build better community links. This is deemed important for several reasons, including the need to:

- Create a self-aware community capable of representation and advocacy.
- Stimulate greater cooperation and learning between related 'knowledge traditions', for example, through the creation of knowledge-sharing platforms.
- Reduce geographical barriers between different regions of the world.
- Provide support to 'lone' FTA efforts in some countries or sectors where little understanding or acknowledgement of FTA might exist.

The FTA seminar series and related efforts of the JRC-IPTS have contributed to these goals, though perhaps not as extensively as some might have wished. A major challenge lies in mobilizing practitioners in community-building activities, with many working to short deadlines on user-led projects and therefore having little time to devote to this. Furthermore, many FTA practitioners, particularly in technology foresight and technology assessment, view themselves as members of a broader science, technology and innovation studies community rather than being part of any broad-based futures studies community. This does not mean that efforts at FTA

[1] See http://www.efmn.eu

[2] See http://www.shapingtomorrow.com

[3] See http://www.prospective-foresight.com

community-building are doomed to fail, only that this context needs to be understood and taken into account.

On a related note, there already exist several groupings, societies, and networks covering technological forecasting, technology foresight, and technology assessment, not to mention the long-standing societies focused around futures studies. There are also academic journals dedicated to the field, as well as a growing library of books. There may be space for new societies and networks, and perhaps even journals, but it would seem prudent to at least survey the current landscape to determine whether the FTA community's needs could be well met by existing forums and publication outlets.

12.3 Raising Awareness of FTA Among Potential Users

Much of this book has been given over to discussion of better integration of FTA into decision-making processes. In particular, the challenge of embedding FTA as part of normal good management practice – without losing the distinctive features of a focus on the longer term and wide stakeholder engagement – have been an important concern. But preceding such considerations is a need to raise awareness of the benefits (and limitations) of FTA among potential users. In other words, how to develop (a) more effective relationships with potential clients, (b) a greater recognition of the value of FTA on the part of government and industry, and (c) an expanding market for FTA services?

Numerous guides have been published over the last few years on organizing foresight exercises, and these have attempted to popularize FTA practice among decision makers. However, these are generally unsuited for introducing FTA practice to the uninitiated and have tended to target those potential users who are already thinking about sponsoring some FTA work. Alternative ideas floated at FTA seminars have included high-visibility publication of successful case studies, the establishment of a PR capacity for the FTA community (which could, for example, prepare articles for the media) and the development of a list of FTA champions linked with a network of FTA users in government and industry. What ever the merits of such ideas – and they would obviously need to be closely assessed – attention needs to be paid as much to the demand side as to the development and 'fine-tuning' of the supply of FTA. This requires not only interpretation and understanding of client needs, but also the shaping of such needs in the first place. This will be a long-term undertaking that is likely to require several different approaches, including, for example, changes in training curricula for decision makers, more and better links between the FTA community and decision makers, and the mobilisation of a society with a heightened interest in the future that demands a similar concern from its decision makers.

12.4 Preparedness to Address Global Problems

Every day issues cry out for systematic application of FTA, yet most debates and actions remain rooted firmly in the present (if not the past), framed as responses to immediate pressing problems and challenges. Many futures studies shift issues to broader, longer term spaces – only to be often dismissed as whimsical and irrelevant. By contrast, many FTA practitioners (for example, working in technology foresight) have served the instrumental needs of public and private decision-making. Whilst this has made them 'useful', the question is whether FTA practitioners could become more proactive in addressing the really big challenges facing global society. It remains unclear how this might be done effectively without major resource investments. Gallant work along these lines is already underway in some parts of the futures studies community (for example, the Millennium Project),[4] but such efforts remain chronically under-funded and rely upon the (largely voluntary) efforts of a few visionary individuals.

Many ideas for a more proactive stance have been floated at FTA seminars – including the formation of FTA key issue taskforces/groups/watchers, the development of a 'rapid response' capability to identify and address emerging issues (for example, through horizon scanning), and the establishment of an ongoing "FTA platform" to analyse/assess existing scenarios and to propose new ones for international debates – but they remain just ideas and have yet to be operationalised. Indeed, the whole notion of FTA intervention on global issues – for example, as a contribution to international forums on key issues – probably needs to be better conceptualised before any specific programmes of work or institutions are created. An important consideration here concerns how the FTA community will be able to garner sufficient credibility, legitimacy and authority to contribute to such global agendas.

12.5 Evaluation, Monitoring and Quality Control

There is some monitoring and evaluation (M&E) going on in and around FTA practices, though it is not yet widespread. Moreover, little existing M&E activity has been collated and codified with the purpose of promoting FTA 'good practice'. In addition, challenges remain in understanding how to incorporate M&E into FTA, on who to involve in M&E, on the choice of indicators and 'measures' of success, around the position of benchmarking and its benefits/limitations, and, in general, on the M&E approach to adopt in FTA. The rationales for M&E, which are manifold and include, for example, issues of accountability and learning, also need to be

---

[4] See http://www.millennium-project.org

better articulated if FTA sponsors and participants are to be convinced of its need. Contributions in this volume should help to provide further understanding around these questions, though more work is undoubtedly needed.

Related to the issue of M&E are questions concerning quality standards in FTA and ethical ‘codes of conduct’. The concept of ‘quality’ in FTA still requires further development and testing, and should undoubtedly refer not only to FTA processes, but also to FTA content (e.g. quality of ‘inputs’ and ‘outputs’) and to FTA impacts. Ethics have been little discussed in FTA, but are perhaps growing in importance as FTA practices expand and have more influence on decision-making processes.

Thus, all in all, there is still a lot to think about and much to discuss. What this volume has tried to do is to provide a base from which these discussions might be anchored. We hope that it achieves this and look forward to the further advancement of the FTA field in subsequent FTA seminars as well as in other fora.

Printing: Krips bv, Meppel, The Netherlands
Binding: Stürtz, Würzburg, Germany